CROP ENVIRONMENTAL ENGINEERING

Prof P. K. Bora

School of Natural Resources Management
College of Post Graduate Studies in Agricultural Sciences, Umiam,
(Central Agricultural University, Imphal)
Meghalaya

Prof B. C. Kusre

Department of Irrigation and Drainage Engineering,
College of Agricultural Engineering and Post Harvest Technology,
(Central Agricultural University, Imphal)
Ranipool, (E) Sikkim - 737135

SCIENTIFIC PUBLISHERS

Published by

SCIENTIFIC PUBLISHERS (INDIA)
5 A, New Pali Road, P.O. Box 91
Jodhpur 342 001 (India)
E-mail: info@scientificpub.com
Website: www.scientificpubonline.com

ISBN : 978-93-90749-46-1 [P/B]
978-93-90749-45-4 [H/B]
eISBN : 978-93-90749-47-8

Printed in India.

Preface

Crop production is function of number of aspects that includes the soil, the surrounding atmosphere and plant itself. Combination of all the three factors results in optimum growth and production. Till few years back all these factors were considered independently. However, now we have knowledge and resources to understand the interactions and its significance.

The subject, **"Crop Environmental Engineering"** was a new subject when we started teaching the Ph. D students of Soil Sciences in a Post-Graduate College of Agriculture in 2009. The students found the subject very alien because this is not exactly discussing the topics that they are familiar with. The same problem we faced when we offered the subject to M. Tech students in an Agricultural Engineering College. As students almost three decades back, we did not have such subjects either. Nevertheless, we found the subjects very interesting because it brings agronomists, soil scientists and soil and water engineers to the same platform. It is an interdisciplinary subject. It tries to look into interactions of agriculture vis-à-vis the environment below and above the ground where the crops grow.

As we progressed, we realized that students still find it difficult to comprehend the subject as there is no single book available which can offer more than two topics. Students need to explore books from soil science to thermodynamics and plant physiology to protected cultivation for this subject. Therefore we have decided to work on this book so that students can at least enter into the subject and get a foundation before exploring deeper into the topics discussed.

No book can be comprehensive. We have tried to discuss as many topics as possible primarily required for this subject. However, we might have left intentionally or unintentionally some topics. We also tried to avoid giving special emphasis to any specialization. Hope this book can be a good guide for the students who try to learn the environmental influences on crop production.

We must admit that we have tried to repackage what is already available in different text or reference books so that the information on each topic is available at the same place. We have tried to make the things very simple to be understood by even a layman with very less scientific background. However, the definitions, equations and deductions of any theory could not be altered and picked up from the sources as mentioned.

We profusely thank our students, past and present, our employer for supporting us to bring out this book.

We are grateful to **M/s Scientific Publishers (India)** for agreeing to bring out this book. Special thanks goes to the management team of Scientific Publishers (India) for their patience with us and the effort made by them for this book.

Prof. P. K. Bora

August 15, 2021 **Prof. B. C. Kusre**

Contents

List of Tables

List of Figures

Chapter - 1

Edaphology and Edaphic Properties

1.1 Soil as a Natural Body

Soil is the outer layer of earth crust formed due to weathering of rocks and other parent materials and a mixture of minerals, organic matter, gases, liquids and innumerable organisms that together supports life on earth. Soil is a natural body also called as pedosphere. It is basically at the interface of lithosphere, hydrosphere and atmosphere. It is also considered as the medium of growth of plants. It can also be considered as the big reservoir of water, organic carbon along with uncountable numbers of micro-organisms. Due to presence of air and water soil is called as three phase body i.e. solid, liquid and gas. Soil matrix and organic matter make the solid phase, the soil-water makes the liquid phase and the gases that fills the interspaces of the solid phase makes the gaseous phase. As it is a habitat of micro-organisms, it also influences the modification of atmosphere.

Soil itself also undergoes continuous modification due to the same weathering agents for which they are formed. Soil erosion, transportation and deposition are processes continuously occurring due to rainfall, runoff and winds. These processes helped in development of fertile valleys and great civilization in the world. However, if these processes go out of balance, the same may become the cause of infertility of the soils and bring hardship to the plant and animal kingdoms.

Soil has multi-facet roles. It gives support not only plants but also the living organisms and the infrastructures created by human being. It is a recycling system of nutrients for plants, a medium for decomposition of organic wastes, regulator of water quality, modifier of the atmospheric system and the provider of suitable eco-system for all. The knowledge of physical, chemical and biological properties of soils are therefore very important. The systematic study of soils, its origin, development and present conditions make an important subject of study almost in all the field of physical sciences. A gram of soil can contain billions of organisms, belonging to thousands of species, mostly microbial. Soil has a mean prokaryotic density of roughly 10^8 organisms per gram of soil.

Study of soils can be classified into two branches: pedology and edaphology. Pedology studies about soil formation, its morphology and classification in its natural forms whereas edaphology deals in soils in relation to the living organisms. The term edaphology though refer to the study of soils, it is more specific to the plant growth and development. It encompasses varieties of factors and soil properties which govern the plant and animal lives in soils. Edaphic factors have been described as those soil conditions that affect organisms living in a particular area. Therefore the study of edaphic factors becomes an important component of ecology too.

When edaphic factors are studied, one needs to consider the edaphic complex which is formed with the vegetations grown or sustain due to soil types, soil physical and chemical properties including soil fertility, soil depth, water holding capacity and drainage class. The functional status of soil ecology is a result of equilibrium of ecosystem where the dynamics of vegetations and its interaction with soil resources are considered. Soil conditions are directly related to the vegetation above ground and vice-versa. Soils often considered as a living body due to the presence of soil micro-organisms.

1.2 Edaphic factors

The edaphic factor pertains to physical, chemical, and biological properties of soil which influence crop growth. Various properties are involved in the process of plant growth which are interrelated influencing the crop growth process in combination and cannot be filtered out to a few. Plants require water, air, nutrients and anchorage from soil. At the same time the photosynthesis process in the plants which produces carbohydrates for plants growth and development requires water and CO_2 apart from sunlight. The sunlight or solar energy is affected by aspects and latitudes which is also closely related to the climate and physiography and other soil characteristics. Broadly edaphic properties can be considered by the way they affect the plant growth and development. They can be classified as the properties influencing the availability of air and water, properties influencing the supply and availability of nutrient and the properties influencing the anchorage of plants to the soil.

The edaphic properties influencing in all the three ways are given in the Table 1.

Table 1.1: Edaphic properties influencing plants growth and development

Soil Properties	Water and Air Movement	Availability of nutrients	Anchorage or impedance to roots
Soil Texture	X		X
Soil Structure	X		X
Depth of Soil	X		X
Depth of impermeable layer	X		X
Infiltration	X		
Hydraulic conductivity	X		
Porosity and Macrospores	X		X
Soil and water potential	X		
Water holding capacity	X		
Drainage class	X		X
Bulk density	X		X
pH		X	
Cation Exchange Capacity		X	
Organic matter content	X	X	X
Nitrate Nitrogen		X	
Organic nitrogen		X	
Total soluble P		X	
Total soluble K		X	
Micro-nutrients		X	
Clay mineralogy	X	X	X
Salinity	X	X	
Heavy metals		X	
Respiration rate		X	
No. of earthworms	X	X	X
Microbial diversity		X	
Composition of organic matter	X	X	X
Root depth	X	X	X

Source: Hillel *et al.,* 2004.

1.2.1 Physical factors

1.2.1.1 Soil texture

Soil texture is the relative proportion of soil separates i.e. sand, silt and clay classified on the basis of their size (diameter). Clays are smallest (<0.002 mm) particles, followed by silts (0.002-0.05 mm), and sands (0.05-2.0 mm). Textures range from those predominantly containing one of the three major particle sizes and thus named after them (clayey, silty, sandy etc.) or various intergradations (silty clay, sandy clay, etc.). Loam is a major textural class which is ideal for plant growth, is not named after a predominant particle class because loam soils have similar amounts of sand, silt and clay. Texture is important for plant growth because it influences water availability as well as soil fertility.Open spaces among soil particles represent the pore space of a soil. This pore space may be filled by water, air, or plant roots. Thus, the amount of pore space greatly influence the amount of water that a soil can contain. The tightness with which water is held in pore spaces determines whether or not the water and dissolved mineral ions will drainthrough the soil profile or remain in place and be available for uptake by plant roots. Coarsetextured soils (e.g., sandy soils) have large pore spaces that do not hold water tightly enough to prevent gravity from pulling that water into deeper soil layers. Very fine textured soils (e.g., clay soils) have very small pores. Small pores hold water strongly and thus retain much water despite the pull of gravity. However, they also hold water against the pulling power of plant roots and so provide only small amounts of plant-available water. Soil texture is also an important determinant of the cation exchange capacity (CEC) of a soil, that is, the ability of a soil to adsorb and exchange mineral ions essential for plant growth. Soils with higher percentages of clay or silt particles generally have a greater CEC.

1.2.1.2 Soil structure

The three-dimensional arrangement of soil particles gives rise to soil structure. In any soils, groups of particles are held together to create lumps of soil materials termed as aggregates. Based on the aggregation soil structure may be spherical (granular/crumby), blocky (angular blocky/sub-angular blocky) and prismatic (columnar/platy). Structural stability is the soil's ability to maintain its structural form despite the application of stresses due to tillage, machinery or rain drop impact. Good soil structure means the presence of aggregation which has positive benefits for plant growth. Favourable soil structure is crucial for successful crop development. The destruction of soil structure may result in a reduction in soil porosity and/ or change to the pore size distribution. In some circumstances a structure less soil mass can result in formation of crusts and pans.

1.2.1.3 Soil depth

Soil depth can greatly influence the types of plants that can grow on the soil. Deeper soils generally can provide more water and nutrients to plants than more shallow soils. Furthermore, most plants rely on soil for mechanical support and this is especially true for tall woody plants (e.g., shrubs, trees). The shallow marginal soils support certain annual plants, whereas deeper soils support herbaceous perennials and still deeper soils are colonized by woody plants. Some soils can develop subsurface hard pans such as duripan, fragipan, etc. that limit the soil depth available to support plants. The net effect of these hard pans is to impede or prevent root growth and thus limit the effective depth of the soil. They also may affect soil oxygenation by restricting drainage at times in which large amounts of water are present. Impermeable layer below the solum may be pre-existing in hilly areas which may effect on the root growth and penetration in to soil. If shallow impermeable layer is present, the crop may not get sufficient anchorage.

1.2.1.4 Soil compaction

Hardness of soil strongly affects root growth. This hardness is usually expressed in terms of penetrometer resistance (the pressure required to push into the soil a cylindrical probe with a conical tip). Root growth slows markedly once this resistance exceeds about 1 MPa and falls away, roughly linearly, to almost nothing at a resistance of about 5 MPa (Materechera *et al.* 1991). The hardness is strongly affected by soil compaction, whether by wheel traffic, the feet of stock, or even the wind induced movement of large tree roots. It is also strongly affected by soil water content, and can vary many fold over the typical range of available water content in the soil (Bengough 1997). Soil hardness may also affect shoot growth (Wolfe *et al.* 1995). Such effects are often explicable in terms of the inability of the hampered roots to supply the shoot with water or nutrients.Leaves typically react very early to uniformly hard soil, even as the first leaf is emerging, when the seedling is still living off nutrients in the seed. Subsequent relative growth rate is much less affected, so that the relative size of plants growing in soft and hardsoil remains roughly constant after the initial response, with consistently smaller leaves and consistently slower rates of leaf appearance with the hard soil (Masle, 1998). Roots may grow into and out of a layer of hard soil, such as a hard pan, and also grow preferentially in cracks or biopores; large continuous pores made by previous roots or earthworms and other soil fauna (Volkmar, 1996). Further, the uptake of water by roots dries the soil and thereby hardens it, making it more difficult for roots to grow in a self-reinforcing process (Bengough, 1997). On the other hand, the exudation of water from root tips may soften the soil in front of the growing root, thereby easing the passage of the root through the soil (McCully, 1995).

1.2.1.5 Porosity

The pore space or porosity of a soil is defined as the ratio of the volume of the pores to the total soil volume. Total porosity is an important soil attribute but the character of the pore space is important for water and air movement as well as root growth and the activity of microorganisms. Pores can be assigned to different size classes according to their significance for different processes. Thus drainage due to gravity will only occur from pores larger than about 30 μm in diameter (macro-pores) while water in pores of less than 0.2 μm is unavailable to plants. Root hairs and larger soil micro-organisms can grow or move into pores larger than 10 μm in diameter. Smaller micro-organisms can exploit pores as small as 1 μm in diameter. Pores larger than about 60 μm diameter can be seen with the naked eye. They, and larger diameter pores are significant for rapid soil drainage and consequently soil aeration. Pores may be planar in shape, completely irregular, or cylindrical. A pore may wide and narrow along its length. Pores may end without connecting to another pore, or be totally unconnected to other pores. Thus parts of the pore system, and the air or water within it, may be unconnected or only poorly connected to the main system (Gardner *et al.* 1999).

1.2.1.6 Soil water

The water content of a field soil can vary from a few percent by volume to more than 50 percent. Water content is measured to establish how much soil water a crop has abstracted and so determine water use efficiencies. However, the water retention and transmission properties of a soil control how much incoming water (rainfall or irrigation) infiltrates, and is then retained in soil and available at a given time for crop usage. Water retention properties are measured in terms of soil water potential. Soil water potential determines whether water is accessible to plants and differences in potential cause water movement from one part of the soil profile to another.Water is present in soils in pore spaces. The saturated water content is determined by the total volume of pore space present. The size of a pore influences how strongly wateris held and how readily water may be transmitted through the soil. The several forces responsible for holding water in soils, including the effect of pore size, are quantified using the concept of potential energy. Water moves in soils and in plants, along potential energy gradients, from zones of high potential to zones of low potential. Water will move into plant roots if the root water potential is less than that of the surrounding soil. The energy with which water is held in soil is quantified through the measurement of soil water potential. The potentials correspond to the soil matrix, gravity, osmotic forces, pneumatic (soil air pressure) and overburden forces. Osmotic potentials are significant for water abstraction by plants in saline soils. It is difficult for plants to uptake soil water held at low matric and/or osmotic potentials, than water held at higher potentials (Gardner *et al.* 1999). Available water of a soil refers to the maximum quantity of water that can be extracted from the soil profile by plants. It is generally defined as the difference in the amounts of water held by a soil when at field capacity and when at its permanent wilting point. Field capacity is the quantity of water remaining in the soil profile about two days after thorough wetting by rainfall or irrigation. The water at field capacity is retained at a matric potential of -0.33 bar. Again, matric potential of soil water at permanent wilting point is -15 bar and plants cannot uptake water at this matric potential. When a soil dries, many changes take place within it. It not only holds water more strongly, but it also gets harder and it transmits solutes less readily so that deficiency of poorly mobile nutrients such as phosphorus is more likely. Plants eventually respond to the falling water potential of their leaves by slowing their growth and closing their stomata. However, plants can also react to a drying soil well before there is a detectable change in the water potential of the leaves. Stomatal conductance (Bates and Hall 1981) and the growth rate of the leaves (Saab and Sharp 1989) may fall, apparently in response to signals received from the roots in the drying soil (Davies and Zhang 1991). Further, both stomatal conductance and rate of leaf growth may fall as the soil dries even if the water status of the leaves is maintained high by pressurizing the roots (Passioura, 1988).

1.2.1.7 Soil Air

The air in the pore spaces of a well-structured, drained soil is composed of about 20% oxygen by volume; this is similar to the amount of oxygen in the atmosphere (i.e. 20.5%). Under anaerobic soil conditions the amount of oxygen may be negligible. In general, the amount of carbon dioxide in soil is about 0.3% which is 10 times higher than the atmosphere (i.e. 0.03%) and this may be as high as 3% under anaerobic conditions. The relative humidity of the soil air is usually close to 100% except in the surface after extended drying. The oxygen and carbon dioxide composition of the soil air fluctuates more than that of the atmosphere. It depends on the rate of use of oxygen & carbon dioxide production by roots & soil organisms and on gaseous exchange between the soil air and the atmosphere. Factors such as the time of year, soil water content, soil temperature and the level of activity of soil organisms, especially the micro-organisms are important in determining the composition of the soil air (Gardner *et al.* 1999).

1.2.1.8 Soil temperature

Soil temperature conditions for optimum plant growth and development vary widely. Decreasing soil temperatures below the optimum level progressively reduce growth rates. Again, Temperatures above the optimum have less of an impact until a maximum is reached when growth ceases. In hot climates, the living root mass of established pasture crops may vary through the season, declining during the hottest months due to high soil temperatures. In many species, the seed must be exposed to temperatures below a certain threshold before germination is triggered. Other soil conditions, such as water content and aeration can exacerbate or ameliorate the impact of sub-optimal temperatures on plant growth(Gardner *et al.* 1999). As with plants, root growth also requires favourable temperature conditions. Better root growth improves the size of the root system and its capacity to provide the above ground parts of the plant with water and nutrients. The rate at which water, and at least some nutrients, are taken up by roots is also influenced by temperature. In those species where the shoot meristem remains at or within the soil surface (cereals in particular), soil temperature in the upper 5 cm directly effects rates of leaf development (Payne and Gregory1988).

Soil temperature influences plants through its impact on soil physical processes, particularly the rate of water loss through evaporation at the soil surface to the atmosphere, and the rate of exchange of oxygen and carbon dioxide between the soil air and the atmosphere. The rate of most chemical soil processes increase with temperature increase, but the relative importance of different chemical pathways may change with temperature. Finally, the activity of soil organisms (macro as well as micro) is influenced by temperature. The role of larger fauna in soil structural development is often severely diminished at low temperatures. As with plants, the activity of soil micro-organisms reaches a maximum at some optimal temperature range but decreases at temperatures above and below that.Organic matter decomposition and mineralization are strongly temperature dependent. Similarly, biodegradation of pesticides and other organic chemicals are also dependent on soil temperature. The activity of crop pests which live in soil is also influenced by soil temperature conditions. For example, populations of nematodes which affect potato production in the sub-tropics, vary in size with soil temperature and crop timing (Greco 1993).

1.3 Chemical factors

1.3.1 Soil organic matter

Organic matter in soils ranges from recognizable plant parts (roots, leaves, stems) to humus, which is partly decomposed plant material that is amorphous and spongy in nature. Organic matter contributes to a soil's ability to retain nutrients and water (i.e., soil's CEC). It aids in holding nutrients because negatively charged compounds in humus attract and hold positively charged plant nutrient ions. It helps provide water because humus can absorb 80– 90% of its weight in water and therefore contributes to a soil's ability to hold water under drought conditions. Organic matter in soils is also an important food source for decomposer and detritivore organisms. Organic matter also contributes to soil color, a factor that affects a soil's thermal properties.

1.3.2 Soil pH

Soil pH is an extremely important ecological parameter. Its most important effect on plant growth is its influence on ion availability in the soil solution. Ions in soils are important for two major reasons. One is that many soil ions contain elements required for plant growth. These elements, called essential nutrients, are primarily obtained from the soil. The second reason is that plants obtain most of their water from soil and the amount of dissolved ions in soil water can influence a plant's ability to take up water. The influence of pH on ion availability stems primarily from the influence of pH on the solubility of the various compounds present in the soil. In general, soil compounds containing some elements are more soluble at some pH values than others. For example, iron is relatively insoluble at pH values of 8 or greater and plants with a high iron requirement may perform poorly in soils with high pH values. Similarly, many heavy metals become increasingly available for plant uptake at pH values of 4–5. Thus, plants growing in low pH soils are more susceptible to heavy metal toxicities.

1.3.3 Cation Exchange Capacity (CEC)

CEC is defined as the ability of the soil to adsorb cations. Many of these nutrients are absorbed in the form of cations. Most soils have at least some ability to hold these ions at negatively charged sites within the soil. The cations are held to the edges of particles within the soil. This is referred to as adsorption. The cations in the soil are divided into acids (eg. H^+ and Al^{3+}) and bases (eg. Ca^{2+}, Mg^{2+}, K^+ and Na^+). CEC is technically defined as the sum of exchangeable bases plus total soil acidity at a specific pH value, usually 7.0 or 8.0. It is usually expressed in cmol(P^+)kg^{-1} or meq per kg of soil. Higher amounts of clay, higher is CEC. 2:1 type of clay (eg. smectites, montmorillonite) have higher CEC than 1:1 type clay (eg. kaolinite). Higher amounts of organic matter mean higher CEC. Amorphous clay minerals and organic matter have a CEC that varies with pH (pH dependent CEC). As pH increases, CEC also increases. For every pH unit above 4.5 there is a 1 cmol (P^+)kg^{-1} increase for each percent organic matter.

1.3.4 Base saturation

Base saturation refers to the percentage of exchange sites (negatively charged site son clay and organic particles) that are occupied with bases (usually $Ca2^+$, Mg_2^+, K+ and Na^+)as opposed to ions that make the soil acid (H^+ or Al_3^+). In technical terms it is the ratio of the quantity of exchangeable bases to the cation exchange capacity. It is expressed in percentage (%). Soils with high base saturations are considered more fertile because many of the "bases" that contribute to it are plant nutrients. Usually the base saturation is 100 per cent when the pH is above about 6.5. Since rainfall tends to leach bases out of the soil, areas with higher rainfall tend to have lower base saturations than areas with lower rainfall, unless the parent material is high in bases, such as limestone.

1.3.5. Essential elements

Essential elements are elements needed by plants for growth and reproduction without which plants cannot complete their life cycle, the function of the element cannot be replaced by another element and the element is directly involved in the plant's growth and reproduction. There are 17 essential nutrients viz. C, H, O (derived from air and water), N, P, K, Ca, Mg, S (mineral nutrients required in large quantity and therefore known as macronutrients), Fe, Mn, Zn, Cu, Ni, B, Mo, Cl, Co (mineral nutrients required in small quantity and therefore known as micronutrients).

1.3.6. Soluble salts

Some ions are abundant in some soils primarily because they have been deposited in those soils in great amounts. Certain salts (often Na, Mg, or Ca salts) may be abundant in some soils in quantities that greatly affect plant growth. These salts may be from sea water or from evaporative concentration of surface water/ groundwater. Extensive irrigation of land inregions where there is high evapo-transpiration can also lead to accumulation of salts on the soil surface. Salty soils (e.g., saline, sodic, saline-sodic soils) can impact plant growth by affecting water uptake, nutrient uptake or by causing toxicity due to specific ion effects. Water uptake can be slowed because the high ion concentration in the soil impedes water movement into plant roots (physiological drought). Nutrient uptake can be affected because ions can competitively inhibit the uptake of essential ions of similar size (e.g., Na vs. K, Mg vs. Ca). Excess ions can also have specific toxic effects on plants by directly inhibiting essential physiological processes.

1.4. Biological factors

1.4.1 Decomposers

Soil micro-flora such as bacteria, actinomycetes (filamentous bacteria), and saprophytic fungi, etc. degrade plant and animal residue, organic compounds, and some pesticides. Bacteria generally, but not exclusively, degrade the more readily decomposed (lower C:N ratio) materials, compared to fungi, which can use more chemically complex materials (NRCS 2004).

1.4.2 Grazers and predators

Soil micro-fauna such as protozoa, mites, nematodes, etc. graze on bacteria or fungi; prey on other species of protozoa and nematodes; or both graze and prey. Grazers and predators release plant-available nutrients as they consume microbes. Often organisms specialize in one type of prey, such as either bacteria or fungi (NRCS, 2004).

1.4.3 Litter transformers

Soil macro-fauna such as earth worms arthropods (invertebrates with jointed legs, including insects, spiders, mites, springtails, centipedes, and millipedes) shred and consume plant litter and other organic matter resulting in increased surface area of the litter accessible to decomposers. Moreover, the organic matter inside their fecal pellets is physically and chemically exposed to microbes living within their guts which break down the plant residue, dung, and fecal pellets consumed along with the soil. These macro-fauna are known as ecosystem engineers that physically change the soil habitat for other organisms by chewingand burrowing through the soil (NRCS, 2004).

1.4.4 Mutualists

Soil is the harbor for mycorrhizal fungi (*Glomus*spp., *Gigaspora*spp., etc.), nitrogen fixing bacteria (*Azotobacter* spp., *Azospirillum* spp., *Rhizobium* spp. etc.) and some free living microbes (*Psedomonus* spp. etc.) which have co-evolved together with plants to form mutually beneficial associations with plants. Mycorrhizae are associations between fungi and plant roots in which the fungus supplies nutrients and perhaps water to the plant, and the plant supplies food to the fungi. These fungi can exist inside (endomycorrhizae) or outside (ectomycorrhizae) the plant root cell wall. The common arbuscular mycorrhizae (AM or VAM fungi) are endomycorrhizae (NRCS, 2004).

1.4.5 Pathogens, parasites, and root feeders

Organisms that cause disease make up a tiny fraction of the organisms in the soil,but it has significant effect on crop growth and development. Disease-causing organisms include certain species of bacteria, fungi, protozoa, nematodes, insects, and mites (NRCS, 2004).

Chapter - 2

Solar Energy in Relation to Plant Growth

2.1 Importance of Solar Energy

Solar energy is vital for plant growth. It is the primary source of energy to maintain the life cycle of any living organism. The significance of plant growth in terms of solar radiation and the efficiency at which it has been used for dry matter production is a well-known fact (Demetriades-Shah *et al.* 1994). The quantity of light i.e. duration and intensity, light composition and quality have profound effect on the plant development and all other physiological activities (Coelho 1980). It basically performs three functions *viz.* photo-energic, photo-cybernetics and photo-destructive. Sunlight is received by the plant kingdom in the form of energy. The solar energy is received by the earth as electro-magnetic radiation consists of particles called photons. Each photon carries certain amount of energy called quantum. These quanta of energy are used and fixed by the plants through the photo-chemical process called photo-synthesis. In photo-synthesis process, 12 quanta of energy are consumed to produce one molecule of carbohydrate, which is the building block of plants. Apart from the photosynthesis, solar energy is necessary to regulate heat energy in the plant body and to maintain water balance. Monteith (1977) stated that efficiency of crop production can be defined as the ratio of energy output (carbohydrate) to solar radiation in thermodynamics term. Apart from solar radiation the availability of moisture and climatic parameters such as temperature also have impact on the efficiency of crop production.

2.2 Energy interaction

Sun is a black body. It radiates the maximum energy across the known radiation spectrum at its temperature. The earth receives the energy from the sun in the range from UV rays to Radio waves. The atmosphere of the earth receives almost a constant radiant energy from 1360 W/m^2 to 1380 w/m^2 (average value being 1370 W/m^{2}called as solar constant) from the Sun. However, due to the process of reflection, diffusion and absorption only about 350 W/m^2 of energy comes to earth surface for utilization by the terrestrial objects. The amount of radiant energy per unit time emitted, received and transmitted is called radiant flux (φ). The amount of flux divided by the area through which radiation flows is called the flux density. The thermal radiation from a hot body is governed by the following laws.

(a) Kirchoff's law

(b) Plank's law

(c) Stefan-Boltzmann Law

(d) Wein's Displacement Law

We shall discuss these laws one by one:

2.2.1 Kirchhoff's Law

Any body at temperature T radiates electromagnetic energy. A perfect blackbody in thermodynamic equilibrium absorbs that all radiation fall on it and at the same time it radiates energy accordingly to its emissive power for the temperature of the body T. This is universal to all back bodies. Kirchhoff's law states that if radiant energy falls on a body in thermodynamic equilibrium, the energy absorbed by the body at the thermodynamic equilibrium is equal to the energy emitted. Mathematically, Kirchhoff's law is expressed as:

$$\alpha_\lambda = \varepsilon_\lambda \quad \text{... (2.1)}$$

Where α is absorptivity and ε is emissivity at wavelength λ.

2.2.2 Planck's law

Planck's law, law of radiation describes the spectral density of electromagnetic radiation emitted by a black body in thermal equilibrium at a given temperature *T*, when there is no net flow of matter or energy between the body and its environment. It is the power emitted per unit area of the body, per unit solid angle of emission, per unit frequency. It is given as:

$$B(v,T) = \frac{2hv^3}{c^2} \frac{1}{\frac{hv}{ek_BT} - 1} \quad \text{... (2.2)}$$

Where, *B*= Spectra radiance of body = 6.62 × 10^{-34} Js.

ν = Frequency of the radiation.

T = absolute temperature of the body.

k_B = Boltzmann's constant = $1.38064852 \times 10^{-23}$ m^2 kg s^{-2} K^{-1}

h = Planck constant.

c = Speed of light in the medium.

Light is emitted in quants and can be considered not only as a wave-like entity but also as a particle, or photon, with the energy given by the Planck-Einstein relation which is given as:

$$E = h\,\nu = hc/\lambda \quad \text{... (2.3)}$$

2.2.3 Stefan-Boltzmann Law

Stefan-Boltzmann law gives the relationship of blackbody radiation to the temperature of the body. It states that emittance of a body is proportional to the fourth power of absolute temperature of the body and written as:

$$E = \sigma T^4 \quad \text{... (2.4)}$$

Where

σ = Stefan-Boltzmann's constant = 5.670374419.. $\times 10^{-8}$ W·m^{-2}·K^{-4}

2.2.4 Wein's Displacement Law

Plank's law can be used to determine the wavelength of maximum radiation at a given temperature T. This law is called Wein's displacement law and given by the following equation:

$$\lambda_{max} = \frac{2897}{T} \quad \text{... (2.5)}$$

Where, T is the absolute temperature of the blackbody.

For example, if the temperature of Sun is considered as 6000K the wavelength of maximum emittance is 0.483 μm.

2.3 Spectral Distribution of Solar Radiation

Solar energy constitutes of wide range of radiation wavelength starting from extremely short wave length with very high content capable of destroying any form of life to very wide wavelength of radio wave with low electro-magnetic energy. The electro-magnetic spectrum of solar radiation is shown in Fig. 1.

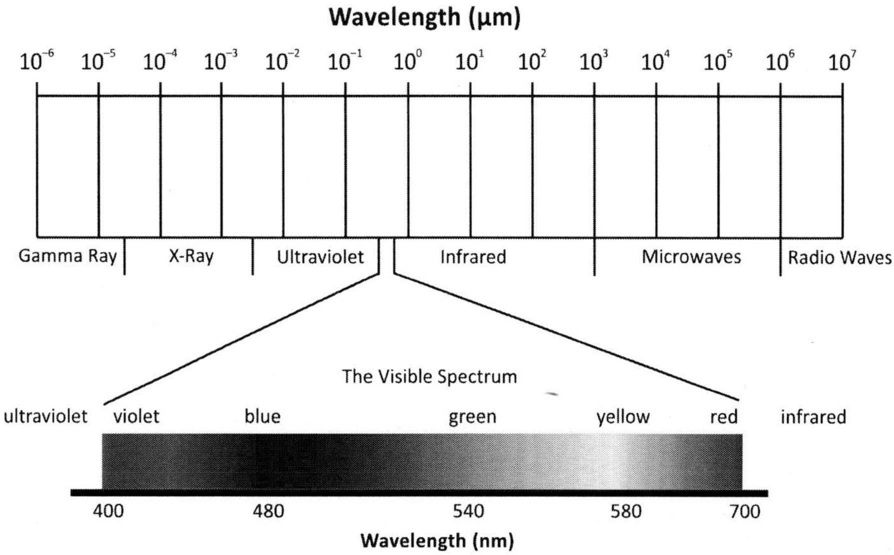

Fig. 2.1: Electromagnetic Spectrum of Solar radiation

The spectrum from 390 nm to 760 nm is the visible portion of solar radiation. This visible portion along with the Near Infrared ray (upto 810 nm) is called Photosynthetically Active Radiation (PAR) useful for plant's photosynthesis process. About 95% of solar energy is contained between 300 nm to 2400 nm. About 1.2 % energy is available in the spectrum below 300 nm and the rest 3.6% energy contains in the spectrum above 2400 nm. Though sun radiates the energy across the entire spectrum, the total energy is modified while entering the earth atmosphere due to reflection, scattering and absorption by the different constituents in the atmosphere. About 31% of total solar energy is scattered back to the space. Atmosphere absorbs about 26% and the rest 43% is absorbed by the earth's surface. Reflected part of the solar energy after falling on any object constitutes the albedo. Albedo is the ratio of outward flux of solar radiation to the inward flux and the albedo of entire earth system is hence 0.31. As a result, earth system is left with around 225 W m^{-2}, which is available for heating the earth-atmosphere system. The broad band solar energy is given in Table 2.

Table 2.1: Broad band of solar energy

Colour	λ (nm)	Irradiance (W m^{-2})	% of solar constant
Ultraviolet	<400	109.81	8.03
Visible	390-770	634.4	46.41
Infrared	>770	634.4	46.40

About 30% of the solar energy is scattered downward. The interactions of solar radiation with the atmosphere result in scattering, absorption and reflection. Two scatterings- one is selective scattering and another is non-selective scattering happen due to interactions of solar energy with the gases and particles present in the atmosphere. In scattering process, the shorter wavelength energy is scattered more severely than the longer wavelength i.e. red and infrared rays. The scattering caused by the gases and particles with smaller diameter than the wavelength is called Rayleigh's scattering. Rayleigh's scattering is the major scattering process where most of the blue colour light is scattered much and as result the colour of the sky becomes blue. The larger particles cause scattering irrespective of wavelength. Mostly white light is scattered in this process which is called Mie Scattering. The atmosphere absorbs about 20% of the radiation. The constituents of the atmosphere that absorb the solar radiation significantly are oxygen, ozone, carbon dioxide, and water vapors.

Earth's atmosphere absorbs about 90% of the outgoing radiation from the earth's surface. Water vapour absorbs in the wavelengths of 5300 to 7700 nm and beyond 2x 10^4 nm. Ozone absorbs in the wavelength of 9400 to 9800 nm. Clouds can absorb energy at all the wavelength. A large part of the radiation absorbed by the atmosphere is sent back to the earth's surface as counter radiation. This counter-radiation prevents the earth's surface from excessive cooling at night. The disposal of solar energy is given in the Table 3.

Table 2.2: Disposal of solar energy

Solar Energy	W m^{-2}
Incidence on the top of the atmosphere	342
Reflected by clouds, atmosphere	77
Reflected from the earth	30
Total reflected	107
Absorbed by atmosphere	67
Absorbed by earth	168
Total absorbed	235

A synoptic day time radiation is given in the Fig. 2.

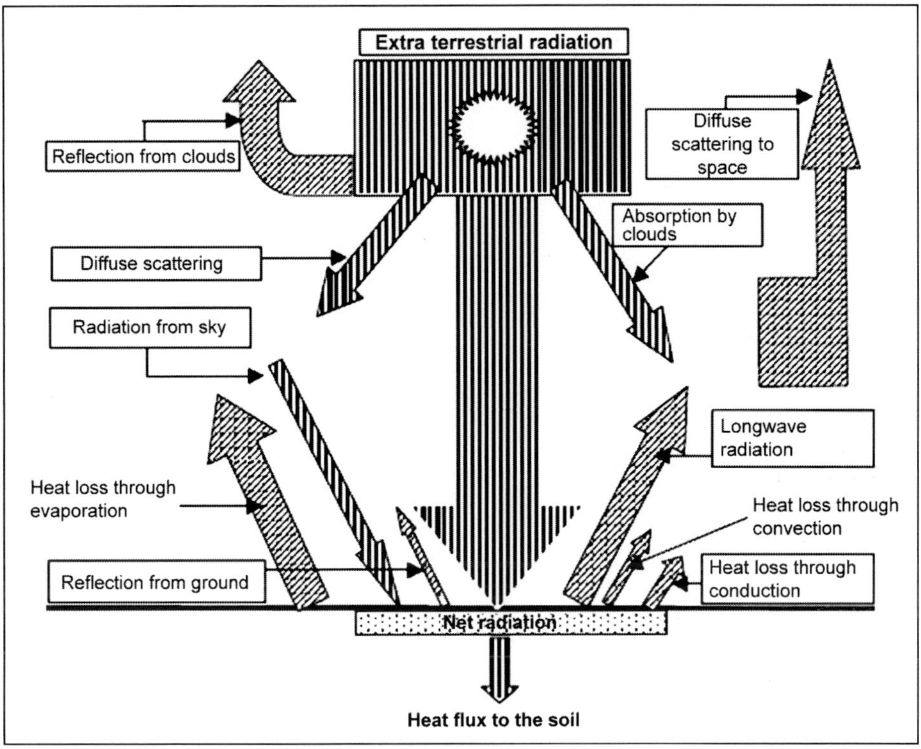

Fig. 2.2: Day time radiation interaction

2.4 Photosynthetically Active Radiation (PAR)

PAR is that portion of sunlight that initiates the process of photosynthesis. Only a small spectrum of solar radiation (0.3 to 0.9 μm) is useful for photosynthesis of the plants. Absorption by plant leaves is very high in the visibile spectrum (0.39 to 0.76 μm), particularly red and blue bands. In the green bands, it is lower and more variable depending on the thickness of leaves. The absorption becomes quite low in the region from 0.7 μm to 1.1 μm and rises again in the infrared band. Absorption in the infrared is about 5% from 0.76 to 1.2 μm and increases to 100% beyond 3 μm. The process of photosynthesis depends on the efficient capture of light quanta by aggregates of pigments which are coloured organic compounds present in the photosynthetic tissues. Most important and abundant tissues participate in photosynthesis are the green pigment chlorophyll. Other pigments such as carotenoid and phycobili also participate in the photosynthesis process. Among the chlorophylls, chlorophyll-a and chlorophyll-b also differ in absorption of spectra for photosynthesis. Chlorophyll-a is found to be prominent at 0.43 to 0.66 μm ad chlorophyll-b is prominent at 0.45 to 0.64 μm. Light energy absorbed by carotenoid and phycobili cannot be directly used for photosynthesis, this energy has to be transferred to chlorophyll-a. However, the pathway of transfer of this energy is not so efficient. Carotenoid is efficient in absorbing 0.38 to 0.52 μm with peak occurring at 0.42 to 0.49. Like carotenoid, phycobili is active between 0.5 to 0.7 μm being peak at 0.6 μm.

Chapter - 3

Environmental Factors in Growth Process

Environmental factors influence the process of crop growth right from germination till the harvest. However, the environmental factors that influence the seeds during the storage are not discussed here. When environmental factors are referred, these are both abiotic and biotic. In this chapter, we shall discuss only the abiotic factors that influence the crop growth process. Among the abiotic factors, we may primarily consider water, temperature and light. These factors act upon the plants in a complex way and it is rather difficult to separate the influence the single factor and its effect on the growth and development of crops.

3.1 Water

The life of plants is much more directly dependent on ***water*** than is life in the animal kingdom. It is because the plants differ from animals. Plants are nutritionally self-sufficient, or autotrophic. Water serves as a hydrogen donor and thereby as a building block for carbohydrates, which are synthesized by plants making use of sunlight. Another inorganic building block used by plants in the synthesis of organic primary products is CO_2, which plants can only take up from the atmosphere at the same time that they return water vapour to the atmosphere. This exchange of gases is necessary because, during their evolution, plants never developed a membrane that was permeable to carbon dioxide but impervious to water vapour. For the exchange of these two gases there are special openings in the leaf epidermis called ***stomata.*** The contra-flow ***gas exchange*** of water vapour and carbon dioxide that takes place between the inside of a leaf and the atmosphere through the stomata is therefore unavoidable. If the exchange of gases is to be maintained for the production of dry matter, growth and development, plants require a continual supply of water in liquid form. This is particularly true for plants other than succulents and halophytes, since the internal store of water is normally very limited relative to the daily loss. The steady use of water demands a constant uptake of water. Plant life depends essentially on water that is stored within the soil and is available for extraction. For extraction of water, plants rely on their root systems, which continue to grow through most of their life. The quality of the soil as a store of water accessible to roots depends on texture and structure. The daily ***throughput of water,*** that is the removal of water from the soil by roots, its movement through the plant in liquid phase, and its final transfer to the atmosphere in the vapour phase, can amount to a considerable quantity in comparison with the mass of the plants involved.

Optimum level of water in plant body, which is directly dependent on the availability of water in the soil, translate into plant growth to the potential and results in good yields. But excess or less water put the plants under stresses and reduces yields. Water as such the most important limiting factor in plant development process. There must be a balance between the water uptake by the plant and the transpiration. If transpiration exceeds the uptake, then the plants experience stress due to water deficit. On the other hand slowing down the root and leaf growth is observed when water logging is encountered by the crops. High positive correlation is observed between water deficit and photosynthesis. Cell development is generally more sensitive to water stress than the CO_2 assimilation and stomatal openings. Mild moisture stress may not affect the photosynthesis but it may after leaf surface area. Leaf area may directly relate to the production of dry matter if the leaf area is limiting the CO_2 assimilation. In growing crop, the lower leaf area index may be quite sensitive to dry matter production. It is also reported that reduction of internode length and smaller leaf are the result of prolonged water stress. Effect of leaf size, rate of leaf expansion and appurtenance of new leaf affect the total dry matter production. The effect of water stress on the crops can be summarized as below:

a. Reduced leaf size and internode length
b. Stunted shoot tops
c. Disproportional root and shoot growth
d. Delayed flowering and fruit setting
e. Reduced seed number, size and seed viability and
f. Halted growth and development and at greater stress, wilting and death of the plants

Excess water can be tolerated by some crops but not by all. In flooded environment the Redox potential is greatly reduced resulting oxygen deficit around the root zone. With the short supply of oxygen the plant metabolism is altered and thereby growth is inhibited. In reduced oxygen environment, the toxic gases are also developed at root zone due to reduction process. Flooded cultivated lands are primarily the reason of production of greenhouse gases from agriculture. The tolerance of flooding varies widely among plant species, cultivars and ecotypes and usually associated with both morphological and physiological adaptations.

3.2 Temperature

Temperature is primarily responsible for regulating the rate of physiological processes and thereby influencing the growth and development of plants. Optimum heat is the basic requirement of life process. Each of the vital processes is dependent on certain temperature ranges. The temperature regime of certain crop within which crops perform well is called cardinal temperature. Cardinal temperature of crop is identified with minimum temperature or base temperature, maximum temperature or the ceiling temperature and the optimum temperature. Base temperature is the temperature below which all the physiological processes stop and there will be no more growth and development. The ceiling temperature is the temperature above which crop will either desiccate or die. The optimum temperature for growth usually occur between 20-25 °C for most of the field crops. Growth rate drops rapidly below 10 °C for C_3 crops, however crop may remain healthy. Growth rate reduces above 25 °C and may cease above 35 °C even with adequate soil moisture. Sub-tropical C_4 crops have optimal temperature between 30 to 35 °C and grow extremely slowly at 10-15 °C. However some exceptions are observed based on local conditions. Generally, active growth of most of the crops is confined between 5-40 °C and varies from species to species. Cardinal temperatures of some of the crops are given in Table 4.

Crop plants are classified into four groups depending upon its physiological behavior towards temperature regimes. They are: cold loving, cold tolerant, heat loving and heat tolerant crops. The cold and heat loving crops may also be divided into four groups. Crops require vernalization, thermo-periodic, non-thermoperiodic and others. The group of crops requires certain degree of chilling or heating at some phases of life cycle for their effective growth and development are grouped as the crops require vernalization. Some crops are very sensitive to diurnal temperature, i.e. day and night temperature variations. These crops are falling under thermo-periodic crops.

Temperature of air alone does not control evapo-transpiration of crops. Vapour pressure gradient is more important in this process, which is determined by relative humidity and wind velocity. It is therefore necessary that air temperature match the evapo-transpiration to maintain internal crop water status.

Saturation deficit may occur if the temperature increases midday that accelerates photosynthesis and fruit ripening. Maximum dry matter is usually produced at temperature range 20-30 °C. It is also evident that higher temperature associated with higher relative humidity becomes suitable for plant diseases. Higher temperature favours the growth of over-ground part of the crops particularly shoot and leaves inhibiting the root development and stolon formation in potato. On the other hand low temperature at night negatively affects plant respiration. The tender leaves of crop plants are also very sensitive to low temperature and frost.

Table 3.1: Cardinal temperature of some important crops

Plant	Cardinal Temperature, °C		
	Minimum	**Optimum**	**Maximum**
Rice	10-12	30-32	36-38
Wheat	3-4.5	25	30-32
Maize	8-12	32-35	40-44
Sorghum	8-10	30-35	40
Barley	3-4.5	20	30-40
Peas	1-2	30	35
Lentils	4-5	30	36
Oats	3-4	25	30
Tobacco	13-14	28	35
Carrot	4-5	8	25

At about 33 °C temperature there is a tendency of equilibrium of temperature between leaves and air. Below this leaves are warmer than the air and vice-versa. If air temperature rises above 33 °C the crops generally suffer from water deficit. At sufficient moisture content of soil the leaves and air temperature do not differ much. Leaves under the shade have higher temperature than the air.

Soil temperature is more important than the air temperature for crop growth and development. When soil temperature goes past 50 °C, all the living tissue will die and at the same time intake of soil water stops at 1 °C. Rice is generally more sensitive to low temperature. Many rice varieties do not germinate if the soil temperature goes below 11 °C. In the tropical region, higher temperature causes degeneration in tubers of potato and it practically becomes absent when soil temperature reaches 29 °C.

As soil temperature increases, the moisture holding capacity of soil also increases primarily due to the expansion of soil pockets. This is more prominent in fine textured soil. Soil moisture movement in vapour phase is important for water supply to the roots and evaporation process. Temperature gradients generally helps in overcoming the resistance due to low vapour pressure gradient in saturated soils. Increased soil temperature may reduce convective flow of air but increase diffusion and hence influence in proper aeration in the soil too. Soil temperature also helps in nutrient uptake to a greater extent as absorption usually increases at higher temperature.

Soil consists of three components *viz.* actual soil mass, water and air. Hence soil temperature depends on thermal characteristics of each of the components. Heat capacity and the thermal conductivity of these materials determine the temperature of soil. On the other hand, these thermal characteristics of the soil mass also depends on the mineral composition, quantity and quality of organic matters, land cover and cover complex, tillage practices, slope and slope aspects of the land, etc. The heat capacity of soil can be defined both on mass basis or volume basis. It is defined as the amount of heat required to raise the temperature of soil 1 m^3 (or 1 kg in mass basis) on by 1 °C.

Volumetric heat capacity is calculated as:

$$C = X_sC_s + X_wC_w + X_aC_a \qquad \text{...... (3.1)}$$

Where the suffices; a= air, w= water, s= solid

Mineral and organic soil heat capacities are about 0.60 and 0.46, respectively

$$C = 0.46\, X_m + 0.60X_o + X_w \qquad \text{...... (3.2)}$$

(Heat capacity for water is 1)

Temperature of the soil depends not only on the quantity of heat transferred but also on the volumetric heat capacity C. Assuming H(z,t) as the heat flux density at depth z and time t, heat storage found from the difference in heat flux density to the heat storage expressed in temperature change (ΔT).

Heat flux H (cal/cm^2/s) is given by:

$$H = -\lambda.\frac{\partial T}{\partial z} \qquad \text{...... (3.3)}$$

Soil temperature varies with depths with higher temperature at the surface during day time and reverse is true at night time. Soil temperature value at any depth from the surface is given by:

$$S_2 = S_1 \exp\left[(z_1 - z_2)\frac{\sqrt{\pi}}{\sqrt{KT}}\right] \quad \text{...... (3.4)}$$

Where, K = thermal diffusivity for dry soil=0.0013

3.2.1 Temperature and Photosynthesis

The role of temperature on net photosynthesis is very important. Photosynthesis undergoes through two reaction phases *viz.* light phase and dark phase. In the light reaction of photosynthesis, adenosine tri-phosphate is formed with energy absorption and transport and in the dark phase of reaction reduction of CO_2 takes place to form carbohydrate. The light phase reaction is chemical in nature is more influenced by the temperature. In the night, if temperature increases considerably, the respiration increases exponentially which negatively affect the photosynthesis process in the dark phase and reduction of CO_2 does not take place effectively. Hence, the role of diurnal temperature is very important to crops and varies with crops and species. Long day temperature is definitely good for higher dry matter production. On the other hand higher diurnal variation of temperature helps growth of thermo-periodic crops in temperate regions. The example of influence of temperature limiting photosynthesis is given in Fig. 3.1.

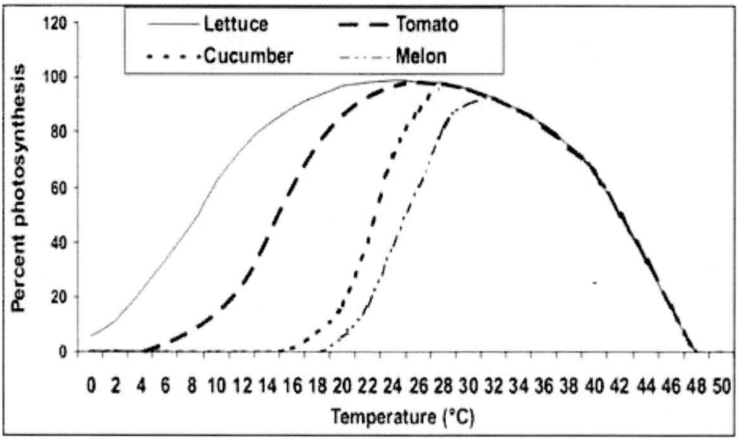

Fig. 3.1: The effect of temperature on photosynthesis process.

3.2.2 Application of Day time temperature in Crop Development Analyses

As the temperature has major influence in the regulation of physiological process in crop development, the temperature or the heat units received by the plants during its life time has major ramification to understand plant growth and development. The concept of heat unit utilized by the crops is referred to as Growing Degree Days (GDD).

Growing Degree Days (GDD), also called heat units, effective heat units, or growth units, are a simple means of relating plant growth, development, and maturity to air temperature. It assumes a linear and direct relationship between the growth of plants and temperature. The growing degree day is calculated as:

Standard degree-day method:

$$GDD = \left[\frac{T_{max} + T_{min}}{2}\right] - T_{base} \quad \text{...... (3.5)}$$

Where, T_{base} is the minimum threshold temperature for a crop.

Growing degree days are also calculated by using maximum temperature instead of daily mean temperature:

$$GDD = (T_{max} - T_{base}) \quad \text{...... (3.6)}$$

Reduced ceiling method:

$$GDD = (T_{max} - T_{base}), \text{ where } T_{max} \leq T_{ceiling}; \text{ or}$$

$$GDD = \sum[(T_{ceiling} - (T_{max} - T_{ceiling})) - T_{base}], \text{ where } T_{max} > T_{ceiling}.$$

If maximum temperature (T_{max}) is greater than the ceiling temperature ($T_{ceiling}$), then set T_{max} equal to $T_{ceiling}$ minus the difference between T_{max} and $T_{ceiling}$.

Even though the concept of GDD is useful in understanding the effect of temperature in the growth and development process of the crops there are certain shortcomings in this concept.

This concept undermines the influence of diurnal or seasonal variation of temperature and the influence of soil temperature in the physiological processes. No differentiation can be made among the different combinations of the seasons. For example, the combination of a warm spring and a cool summer cannot be differentiated from a cold spring and a hot summer. No allowance is also made for threshold temperature changes with the advancing stage of crop development. The net responses of plant growth and development are to the temperature of the plant parts themselves, and they may be quite different from temperatures measured in a Stevenson's screen. Though this difference at a particular time may be small, the cumulative effects through an entire growing period can be very large. The effects of topography, altitude, and latitude on crop growth cannot be taken into account. Wind, hail, insects, and diseases may influence the heat units, but these cannot be accounted for in this concept. Soil fertility is also found to influence on crop maturity which cannot be explained in this concept.

Apart from GDD, some other derived heat units are also in use. These units use sunshine hours Helio Thermal Unit (HTU) and actual day length (Photo Thermal Unit) in combination with GDD to rationalize the effect of actual radiation and day length without giving emphasis only on the maximum or minimum temperature.

3.3 Light

Light provides the energy to complete the photosynthesis process in plants. It is not only the quantum of light the plant receives, light quality, duration and intensity are equally important for plant growth process. In the previous chapter, it has been already discussed, how the solar energy is attenuated or altered through the process of scattering, diffusion and absorption by the atmospheric gases.

3.3.1 Quality of light

The light energy received by the plants in the normal process after the atmospheric interference generally very good for plant development process. The sunlight received by the earth surface consists of ultraviolet rays (<0.39 μm), visible rays (0.39 μm to 0.76 μm), near infrared rays (076 μmto 0.92 μm) and the infrared rays (0.92 μm to 1.2 μm). Infrared rays are generally necessary for maintaining heat balance in the plants. The near infrared rays are responsible for plant elongation, seed germination, fruit coloration, flowering and photoperiodism of plant. The regions in the solar spectrum from 0.6 to 0.7 μm and 0.4 to 0.5 μm are concerned with photosynthesis and formative process of plants. Radiation received in the range from 0.3 μm to 0.4 μm produces formative effects leading to shorter plants and thick leaves.

The latitude, altitude and atmospheric conditions alter the spectral composition and distribution. At higher altitude, usually the percentage of ultraviolet light increases and the visible spectrum is reduced. Cloud cover and shade also change the composition. Under the shades, the percentage of infrared rays is more. In the cloudy days, the percentage of ultraviolet light and infrared rays are reduced considerably. The light composition and intensity can be manipulated to a greater extent under the controlled environment agriculture to use the light energy in effective ways for increasing productivity and quality of the produce.

3.3.2 Duration of light

Plants behave conspicuously based on the actual photoperiod or day length. Based on the flowering behavior of plants towards the duration of sun light, they are classified as long day, short day and day neutral crops. This means continuous period of dark period in a day. Besides the three groups of plants, another group of plants is also identified as intermediate type which do not flower when the days are less than 12 hours and more than 14 hours. Again the long and short duration crops may be sub-classified facultative or strict types. In facultative type, the effect of photoperiod is additive and the plant will flower in a subsequently unfavourable period of light. In the strict type flowering will not occur if the day length differs from the critical values.

In addition to flowering, the photoperiod may inculcate or break the bud dormancy and winter hardiness and also may influence in elongation and branching of stems. It is also found to influence on shape, size, succulence and abscission of leaves. Photoperiod is equally responsible for fruit setting, sex expression, germination and formation of tubers and bulbs.

3.3.3 Light Intensities

Light intensities have also considerable influences on the performances of crops. Depending upon the varying degree of influence of light intensities on crop performances the crops are classified into heliophytes (sun loving type) and sciophytes (shade loving tpe). They can again be classified as facultative and obligative types. Facultative types of plant do not strictly adhere to the intensities requirement but can perform relatively good when the light intensity regimes are changed.

The influences of light intensities on crop photosynthesis are studied through two light intensity points. One is called 'compensation point'. It is the minimum intensity of light at which the rate of respiration equals the rate of photosynthesis. The higher the temperature the higher is the compensation point. Compensation point can vary from about 32 lux to 4300 lux. At normal air temperature, a halophytes plant can have compensation point at 540 lux. Another point is called 'saturation light intensity'. It is the minimum intensity of light at which the rate of photosynthesis reaches the maximum values. Most of crops reach saturation point at 1/5 of the full sunlight. The sciophytes can have saturation point at about 10 klux.

Light intensity is one of the most critical variable influencing crop production through its effect on overall photosynthesis process, leaf area ration while the photoperiod may exert its effect on crops through the length of growing phases as well as shape, size, area and retention of leaves. The light intensities and photoperiod vis-à-vis crops also determine the crops and species distribution in the nature and over all bio-diversity.

Chapter - 4

Modification of Environmental Factor for Plant Growth

4.1 Introduction

In Chapter-3, it has been discussed how the environmental factors *viz.* water, temperature and light influence the crop production. Crops are selected by nature. The native crops of any place successfully adapt to the local environment and accordingly perform well. These crops are often found to be resistant to different biotic and abiotic stresses because they adapt to the environment and become naturally resistant to different stresses. But crop is not about native species and varieties alone. Higher productivity and profitability is the hallmark of modern agriculture. New varieties are developed; exotic species are adapted to new environment. Sometimes, these crops or varieties need to compete with the plants in the adopted environment. Such situation requires certain modifications of the growing environment. Sometimes, stress of some kind can also be mitigated for better crop performance. These often need external support. Instead of growing crops only in the natural situation, modification of growing environment which may in other words called as engineered environment may help in increasing crop productivity and profitability to a great extent. Lack of optimal environment often leads to unnecessary stress to the crops and reduced productivity. With the help of engineered environment, the optimal micro-climate to the crops can be provided without compromising the yields.

Apart from common stresses related to crop environment such as bad weather, insufficient water or unfavourable temperature regime and pest and diseases, farmers often encounter with the extreme situations. Drought, flood, heat wave, frost, cold wave, hail storms, cyclone, etc. may be devastating. The climate change is making such situation further uncertain.

These extremes can devastate farmers by damaging not only the crops of the season but also the entire agricultural ecosystem. Global warming, erratic rainfall, frequent drought, etc. are also inviting many other associated problems which are not known to the farming community. Even though extreme weather may require extreme measures, certain measures can be easily taken by the farmers, specially modification of cultural practices through which the micro-climate of cropping environment can be modified to a certain extent to avoid the damages due to unfavourable conditions. In this chapter we shall discuss about the modification of crop growing environment for enhancing profitability of agricultural practices. Different crops have specific climatic requirements at different growth stages. Thus, raising crops under favourable microclimate as per requirements of the crops needs different management techniques so that sensitive growth stages of the crops can coincide with the favourable micro-environment.

4.2 Forms of environmental modifications

The modification of environment can be of two types: long term modification and short term modification. For short term strategies mostly cultural practices and adjustment are made. Soil moisture conservation, alteration of radiation interception, temperature moderation, etc. are some of the practices one can adopt for improving crop productivity. Some of the common practices are listed below:

i. Selection of appropriate crops, species and varieties.
ii. Alteration of sowing dates to escape the stress
iii. Changing of plant population and field orientation.
iv. Crop rotation
v. Tillage practices
vi. Planting methods
vii. Inter-cropping
viii. Soil moisture conservation
ix. Shelter belts or wind breaks, etc.

All the agronomic practices viz. sowing time, planting methods, plant populations, inter cropping, soil manipulation techniques, moisture conservation techniques through mulching, etc. affect the overall crop environment which includes light and air availability, canopy temperature, advection, evapo-transpiration, diffused light, reflected energy from the ground, etc. These all can be easily done at farm level without major alteration at production process yet one can achieve a great benefit.

Long-term strategies include major structural changes of farm production systems and need careful agro economic planning and realization at societal level; these adaptations also involve a set of sectors and stakeholders, such as policy, research, water and land planning. Some examples of long-term adjustments are changes in land use and landscape structure, breeding and biotechnology applications, crop substitution and changes in the farm production type. A change in the landscape structure may include the soil and water conservation works such as terracing, bunding, water harvesting and bio-engineering works for restoration of soil and arrest the rate of degradation, etc. Change of slope, slope aspects for modifying radiation incidence on crop can also be thought of. Microclimatic modifications can be an effective adaptation strategy of technological management for reducing the climatic risks and better utilization of available natural resources result in higher agricultural productivity and thereby enhancing food and livelihood security of small and marginal farmers.

4.2.1 Selection of appropriate crops, species and varieties

Every crop has its own unique requirement of environment. A warm climate crop obviously cannot be cultivated in temperate region. Thermal requirement, day length, photo-period, etc. are different to different crops. Hence, while selecting non-traditional crops or variety one needs to study the requirements carefully. Breeding programmes can also be planned for variety development suitable for the environment. Varieties for drought tolerance, cold tolerance, acid tolerance, high uptake of certain nutrient, fortified varieties may be considered under this strategy. Diseases and pests resistant or tolerant varieties are often commonly included in the breeding programme. The climate change scenario has made such variety development programme more important to adapt the inevitable changes. Changed climate also often select the species as species migration is often a common problem in global warming period. Varieties earlier cultivated in the warm region are gradually making inroads to the temperate region. In such cases, the variety selection of the new crops becomes mandatory research issue for agronomists as well as plant breeders.

4.2.2 Alteration of sowing dates to escape the stress

Sowing time is an important production component that can be manipulated to counter the adverse effects of environmental stress. This is accomplished through shifting of sowing time so that any stress caused by environment is avoided during the critical stages of crop growth. Matching the phenology of the crop to the duration of favourable environmental conditions by selecting the most appropriate sowing time to avoid the periods of stress is crucial for obtaining maximum yields under changing climate. Favourable microclimate can be created by modifying the date of sowing, which takes care of the deleterious effect of high temperature during reproductive growth period. Temperature is one of the most important elements of climate which determines directly the potential productivity level particularly for winter crops. Date of sowing influences the yield considerably and delays in sowing subject the crop to mature early due to rise in temperature resulting in decreasing the number and size of grains. Thus, sowing time has a significant effect on heat and radiation use efficiency of crops and it can be adjusted to enhance the input use efficiency in crop production

4.2.3 Changing of plant population and field orientation

Row spacing has a great impact on canopy temperature and radiation interception by the crop. If row spacing is more, radiation interception by the crop decreases, more radiation falls on the soil surface and increases the soil temperature. On the other hand, if row spacing is less, radiation interception by the crop increases and its transmission towards soil surface decreases, which decreases the soil temperature. Shade has a direct effect on soil temperature, and moisture loss from evaporation and transpiration. By reducing incoming radiation, soil moisture is lost at a lower rate. Reduced air temperature can benefit crops that are sensitive to heat, such as tea, coffee and cocoa. Plant architecture changes affect the canopy light environment, yield and yield components of narrow-row cotton. Alterations of canopy architecture that allows more light penetration into the lower depths of the canopy may be a way to increase the yield of cotton through greater boll production. Appropriate row spacing and orientation help in trapping of radiation by the canopy, thus leading to higher radiation interception and crop productivity. It is also observed higher light interception in wheat and barley along east-west row orientation as compared to north-south. Hence, the row spacing and orientation significantly affects the crop productivity under specific conditions required for specific crops.

4.2.4 Crop rotation

Crop rotation is growing of different crops in sequence in the same cultivated area across the sequence of cropping season. Crops are often classified crudely as soil building crops, soil maintaining crops and soil depleting crops. Certain crops and varieties are very exhaustive and have higher demand for nutrients and water and other inputs. Such crops often lead to soil degradation both chemically and biologically. Some crops can fix atmospheric nitrogen and improve soil health. One needs to mix both types of crops in ones annual crop calendar so that sustainable production can be achieved. At the same time one should not grow the same sequence of crops over a period. Same crop sequence growing in the same area may create favourable environment to certain pool of pests and diseases which will be difficult to evade. By changing crop sequence with mixing of soil building and soil depleting crops one can help in mitigating endemic pests and diseases and maintain the soil health. For example, growing rice-rice, maize-potato, maize-rice, etc. in sequences for long period will definitely damage the soil fertility as no crop in the sequences is soil building crop. It is therefore necessary to incorporate green manure, pulses or such soil building crops in the sequences.

4.2.5 Tillage practices

Tillage is defined as physical manipulation of soil, which is intended to destroy weeds, incorporate crop residues and amendments into soil, increase infiltration and reduce evaporation, prepare seed bed and break hard layer to facilitate root penetration. Tillage practices modify the crop microclimate by affecting the soil properties. Tillage increases the porosity and decreases the thermal conductivity of the soil and increases the soil temperature. The upper loose soil surface also acts as mulch and conserves soil moisture by decreasing evaporation losses. At the same time tillage leads to soil erosion, oxidation of soil organic matter and enhances the loss of C by way of CO_2. Tillage also needs excessive energy input in the farm sector particularly fossil fuel if machiney are used. Tracking of heavy machinery during tillage causes plough sol formation and compaction of soil reducing soil infiltration, poor germination and water stagnation. Excessive tillage particularly in the wetland situation the soil structure is greatly damaged through tillage. Excessive tillage to achieve good tilth for sowing of seeds breaks the stable aggregates in the soil. Soil aggregates resist to soil erosion and help in water conservation through increased infiltration. Tillage is hence considered responsible for excessive loss of top soil and fertility due to both wind and water erosion. Modification of tillage is therefore considered as a tool to improve sustainability of soil fertility. Tillage practices like conservation tillage, where 30% of crop residues are essentially kept over the surface of the soil to reduce erosion, is therefore considered as a good step for soil sustainability and climate protection. Zero tillage is extreme form of conservation tillage. New age machinery used for tillage is also modified. Sharp edge, light weight implements can achieve all the advantages of tillage and at the same time mitigate the problems like soil compaction, structural damage of soil and lower draft power requirement. Conservation tillage also helps in decreasing in morning soil temperature by 0.1 to 0.4 °C and increase in crop yield.

4.2.6 Planting methods

Method of planting also affects the crop microclimate due to its effect on soil properties. Bed planting enhances 10% yield of wheat due to improved soil physical conditions as compared to flat sowing. About 26-42% of water can be saved through beds as compared to flat sowing in various crops. Bed sowing produced significantly taller plants with greater biomass and grain yields probably due to better nutrient availability, good soil conditions and weed control in beds. Furrow seeding improves the photosynthesis and dry matter production in cotton. It is also observed that higher canopy temperature and PAR interception is achieved in bed planted rice as compared to conventional planting. Direct seeding of rice is found to be better in many ways over transplanted rice in terms of environmental sustainability by means of energy saving. Technique of direct seeding of rice proved to be beneficial in post flood scenario when transplanting may take longer time and delay in overall crop production process. Zero tillage seeding is another technique of modified planting of seeds which has been proved to be beneficial for most of the crops particularly in the situation of post rice rabi crops cultivation. Zero tillage seeding does not require land preparation and hence farmers can reduce the turnaround time of next season crop production and use the suitable cropping season.

4.2.7 Inter-cropping

Intercropping means growing of two or more crops simultaneously on the same field such that the period of overlap is long enough to include the vegetative stage, has significant effect on microclimate and resource use efficiency. Intercropping, double cropping and other mixed cropping practices lead to enhanced efficiency of farm resources with sustainable crop production. Intercropping increases the radiation interception by the canopy and decreases its transmission towards the soil surface, thus also decreasing the soil temperature. Improved productivity thus results from either greater interception of solar radiation or higher light use efficiency or their combination. Crop selection is however very important in inter-cropping. The crops grown together must be symbiotic rather than competitive for nutrient, light interception and water consumption. The main objective of inter-cropping is to improve better resource utilization and to achieve higher productivity from the same area of land.

Efficient radiation harvesting in intercropping of maize and beans as compared to sole crops are reported in different literature. Higher PAR interception and radiation use efficiency in maize and soybean intercropping as compared to sole maize crop. Intercropping, modifies the microclimate within canopy and can moderate temperature extremes, lower temperatures with reduced air movement leading to decreased evaporation rates and increased relative humidity, which is important in avoiding desiccation and providing favourable growth conditions even during the periods of moisture deficit.

4.2.8 Soil moisture conservation

Improving water holding capacity of soil is one method of moisture conservation. Increased infiltration in the soil also leads to reduced soil erosion by way of reducing erosive runoff. Improving organic matter in the soil, addition of soil amendments for improving water holding capacity are common practices adopted by farmers. Biochar is a material used by farmers to improve water holding capacity of soil. Modification of land configuration also helps in moisture conservation. Conservation terrace, contour trenching, etc. can improve moisture availability.

Mulching is another technique to modify micro-climate apart from excessive evaporation from the soil surface. Mulch kept on or applied to the soil modifies crop microclimate above and within the soil as well as above and within the mulch. Residue retention, mulch application and/or live mulch planting are few methods followed by farmers throughout the world. These mulches modify crop microclimatic aspects, viz. solar radiation, reflection and absorption, shading, thermal radiation, temperature, humidity, wind/ air movement, evaporation (crop and soil), soil moisture (surface and within) and crop composition, structure and growth. They have marked effects on the soil-residue-atmosphere interfaces, depending on the characteristics of the residues, soil moisture and temperature alterations simultaneously, and modifications in the plant physiological functions due to environmental conditions created by mulches. Mulching produces new microclimates in the original soil, in each new layer thus created and in the layer of air that covers the new surface. For example, the transmittance of global solar radiation through mulch of crop residues must be considered to understand the influence of surface material on the energy balance and temperature of the underlying soil and the overlaying material and air in conservation tillage systems. Zero tillage with surface mulch has been found to be a promising alternative to conventional tillage and gave higher yield as a consequence of improved water conservation. One of the major advantages associated with straw mulching is retaining greater availability of soil water by controlling evaporation loss from the soil surface and improving water infiltration and also protect from wind and water erosion. Mulching may be positive for the seedbed, yet thick layers may prevent germination of seeds that positively respond to light and to temperature fluctuations and can deter establishment and performance of shaded emerged seedlings. Traditionally, over the ages unconsolidated organic materials such as bark or straw were used as mulches/crop residues for modifying the soil thermal properties. In another, narrower definition, a mulch is 'a protective layer of organic or inorganic material left on or near the soil surface as an aid in stabilizing the surface and improving soil microclimate conditions for establishing vegetation'. Mulch obstructs the solar radiation reaching the soil surface. They increase the soil temperature during winter and decrease during summer season. White and light coloured materials which increase the reflectivity, are used to decrease soil temperature during summer season, whereas, dark-coloured materials and black plastic mulches are used to increase soil temperature during winter season. Straw mulches can be used during both the seasons as they increase soil temperature during winter season and decrease during summer season. Due to obstruction of the exchange of water vapours, mulches decrease the rate of evaporation from the soil and conserve soil moisture for use by the plants. Straw mulch improves the crop productivity through optimizing the hydrothermal regimes of soil. The benefits are more in summer/*kharif* season and on soils having low retention. It may be positive for the seedbed by preventing water loss from the soil surface; yet thick layers may prevent germination of seeds that positively respond to light and to temperature fluctuations. Hence, by the various methods of water conservation, prevention of erosion, weeds management, reduces the soil temperature in summer and increasing during winter season, etc. mulching helps in increasing the crop productivity.

4.2.9 Shelter belts or wind breaks

Trees that are used as windbreak protect a field from prevailing wind patterns, greatly lowering the wind speed before reaching crops. The temperature behind the shelterbelt is usually slightly higher, as the cooling effect from wind no longer applies. Humidity is higher as well, as it is no longer transported away. Reduced exposure to wind and increased humidity reduces evapo-transpiration rates from both soil and crops and can increase water use efficiency.

The main function of shelterbelts is to reduce the wind speed and cold advection. Due to reduction in wind speed, they decrease the evapo-transpiration and increase the temperature and moisture. Windbreaks modify the microclimate by providing the shelter, which reduces the wind speed and turbulence. Wind breaks protect the plants from the high intensity winds. They modify the water table by reducing the water losses and hence they improve the crop productivity. Wind breaks also reduce the wind erosion.

4.3 Achievement through short term modifications

4.3.1 Controlling heat load

Changes in the soil temperature can be induced by covering the surface of the soil. Growing a cover crop is one well-recognized method to do so, and brings additional positive effects related to organic matter content of the soil, seed germination and moisture conservation. Trees and other tall plants that create a canopy (over-storey) that covers other plants greatly modify temperature conditions under the canopy. Shade from the canopy reduces solar gain at the surface of the soil, and helps the soil to retain moisture.

Heat load on crop canopy can be increased (during winter) called heat trapping, or decreased (during summer) called heat evasion. Heat evasion can be achieved by shading the crop, irrigation management or by using anti-transpirants, whereas heat trapping can be achieved by adopting appropriate row direction or plant density.

4.3.2 Water balance

It involves in increasing the amount of water stored in the root zone (by strip cropping, contour cropping, ploughing, terracing or bunding), increasing infiltration (by increasing tillage or increasing row spacing), reducing soil evaporation (by using mulches and crop residues, plastic covers or thick cultivation crops), modifying transpiration (by using antitranspirants, substances forming film on leaf surface, stomata closing materials and reflectants). Whereas, winds can be controlled by applying wind breaks or shelter belts. These protections reduce wind speed on the leeward side and save the plants from freezing injury and mechanical damage.

4.3.3 Solar Radiation

Modification of solar radiation can be accomplished by increasing the surface absorptive power, reflective power of the surrounding objects, exposure through site selection, increasing the radiant energy by fog dissipation and by adopting appropriate row direction. Similarly, soil temperature can be modified by mulch application and soil tillage etc. Mulch helps in reducing the soil temperature in summer and increasing during winter season. Similarly soil tillage by harrowing or inter-culture operations also help in modifying the soil thermal regime. Similarly, protected cultivation under controlled environment like greenhouses (for temperature, humidity and CO_2 control), poly houses (naturally ventilated or with controlled environment) and shade houses (for crop production in warm climate) provide other means for raising crops under controlled microclimatic conditions.

4.4 Long term strategies

It has been already discussed that long term modifications of agro-environment includes changing of land configuration, change of breeding and biotechnological programmes related to crop development as well as policy change. Government mission mode programmes often change the entire agro-environment through certain economic and policy interventions.

Government missions like 'Pulse Mission', 'Oil Seeds Mission', etc. modifies the agricultural production process in long term perspectives. Protected agriculture, precision farming and organic farming are few of the new paradigms need log term planning and policy interventions. Protected agriculture can be termed as controlled environment agriculture where all the environmental factors can be controlled depending upon the crop requirement. Optimum temperature, humidity, light, CO_2 all can be supplied in controlled environment to achieve potential yield and best quality of yield of any crops. Such controlled environment agriculture naturally comes along precision management and automation. Best practice comes at a cost. Investment is also very high in such agricultural practices in modified environment. Hence, a conducive policy regime is necessary for adopting these long term modifications.

Chapter - 5

Different Stress on Crop Growth

It is a well-known fact that water is essential input in productivity and production of plants and vegetation. The plants grown in soils, extracts water and minerals through its root system. The minerals present in the soils are not directly available to the plants but in the form of solutions. The shortage of water supply leads to affecting the rate of growth to no growth ending up in wilting or death of plants. Many a times water is available in the soil but plants is starved of it. It is due to the forces that holds water with the soil particles. When the intensity of these forces increases the growth of plant is inhibited and is termed as soil moisture stress.

5.1 The mechanics of water uptake

Why water is so important for plant growth and what is the role of water in the process. The knowledge of this aspect will help in planning water resources with respect to irrigation or agricultural water management. As stated earlier plant takes nutrient in the form of solution only. The water enters in the plant through fine root system that consists of root hairs at their tips (Fig. 5.1). The water then passes through various cells before passing through endodermis that controls entry of water in to group of cells known as xylem. The endodermis is a single layer of cells that borders the cortex of a root. It regulates the entry of desired nutrients through the roots into the vascular core (Rost *et al.* 2006). The xylem are vascular tissues and acts as a tubular system to transport water upwards in the plant to the leaves (Roberts 2007). In this process the nutrients reaches the leaves where the process of photosynthesis is completed. In the leaves exists stomata from where the water leaves the plant system to atmosphere. Stomata are small pores (Fig. 5.2) on the surface of leaves that controls gas exchange (CO_2 and O_2) and water vapor between the plant surface and atmosphere (Hetherington &

Woodward 2003). The process of exchange of water with the atmosphere is generally known as transpiration. The process of transpiration leads to creation of suction that makes the roots to absorb water from the soil for replenishment and thus the processes goes on. The basic property of water is to move from high potential to low potential. At the root hairs the potential difference is created that makes the water to move from soil to the plant bodies. The driving force is dependent on the energy by which the water is held by soil. The movement of water in soil is basically controlled by water potential and gravitational potential. The movement of water in soil is better described by Darcy's law. However, the water movement in the soil under saturated and unsaturated conditions also differs. Many studies have indicated that change in soil water potential adversely impacts the physiological development of plant such as root proliferation and thus impacts the water availability leading to reduced stomatal conductance and photosynthesis and transpiration. The reduced metabolic activities are directly related to its yield. Therefore, it is necessary to understand the movement for achieving better water management.

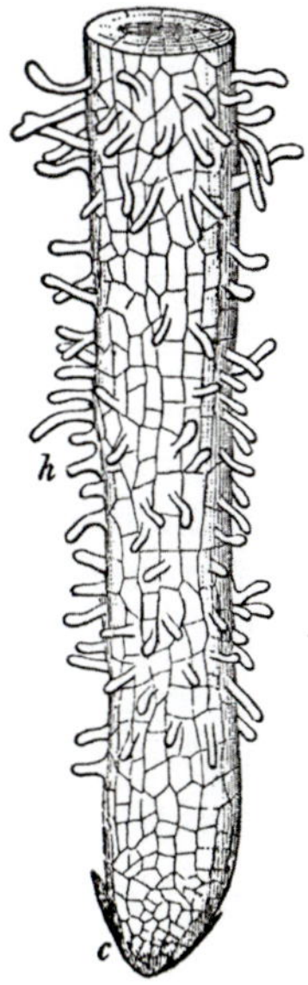

Fig 5.1: Figure of typical root hair

Fig. 5.2: Image of stomata generated by compound microscope

5.2 Soil Water

The water present is the soil is held by different forces. The forces that holds the water determines the rate of entry and retention in the soil. This phenomenon is mainly based on the physical properties of the soil. The soil basically consists of mineral particle, organic matter, living organisms, gas, and water. The mineral particles that forms the bulk of soil are classified into three types based on the size viz., sand, silt and clay. These mineral particles are formed due to weathering of parent rocks. The mineral particle in the size range of 2 cm to 2 mm are categorized as gravel, and the particles size below 2 mm are known as *fine earth.* The fine earth is the main component that controls the retention and movement of water in the soils. The most commonly used classification of soil are proposed by United States Department of Agriculture (USDA) and International Soil Science Society (ISSS) which as shown in Table 5.1.

Table 5.1: Soil classification based on particle size

Fraction	Particle diameter	
	USDA	ISSS
Gravel	> 2 mm	> 2 mm
Very coarse sand	1 to 2 mm	-
Coarse sand	0.5- 1 mm	0.2 to 2 mm
Medium sand	0.25 to 0.5 mm	-
Fine sand	0.1 to 0.25 mm	0.02 to 0.2 mm
Very fine sand	0.05 to 0.1 mm	-
Silt	0.002 to 0.05 mm	0.002 to 0.02 mm
Clay	< 0.002 mm	< 0.002 mm

Sand and silt particles are normally spherical or cubical in shape. These particles are simple in nature with large volume of pores in between which makes movement of water within the soil very easy. These large volumes of pores are generally known as non-capillary pores. The soil particles are loosely held and are non-cohesive in nature. Such soils have low water holding capacity. Unlike the sand and silt particles the clay are particles are complex in shape. These particles can be platy in nature or lath shaped depending on the structure of specific types of clay. The irregular shape of the clay particles enhances the surface area that enables to hold more water. Although it holds more water the but the colloidal size and the negatively charged ions holds the water tightly restricting its movement within the soil. The negatively charged clay particles control the ability to hold the cations.

Understanding the physical properties of soil and its relation with soil water movement can improve water management practices. Soil texture, structure, porosity is some of the physical properties that have influence on infiltration, permeability and water retention capacity of any soil. The relative concentration of sand silt clay determines the texture of soil. The texture has a considerable influence on the soil and air phase of soil. The arrangement of the individual particles into aggregates decides the structure. The aggregates can range from loose and friable to have distinct pattern. Fig. 5.3 shows the soil textural triangle that decides the type of soil based on particle size distribution. The inter particle space in a soil determines the porosity. The porosity is outcome of the soil texture and structure. The porosity indicates the volume of water that can be stored in the soil. Table 5.2 gives the water holding capacity of different soils:

Table 5.2: Available water capacity by soil texture

Textural class	Available water capacity (cm/m depth of soil)
Coarse Sand	20.8- 62.5
Fine Sand	62.5- 83.2
Loamy sand	91.7- 100.0
Sandy loam	104.2-116.7
Fine sandy loam	125.0-166.7
Silt loam	166.7-208.3
Silty clay loam	150.0-166.7
Silty clay	125.0-141.7
Clay	100.0-125.0

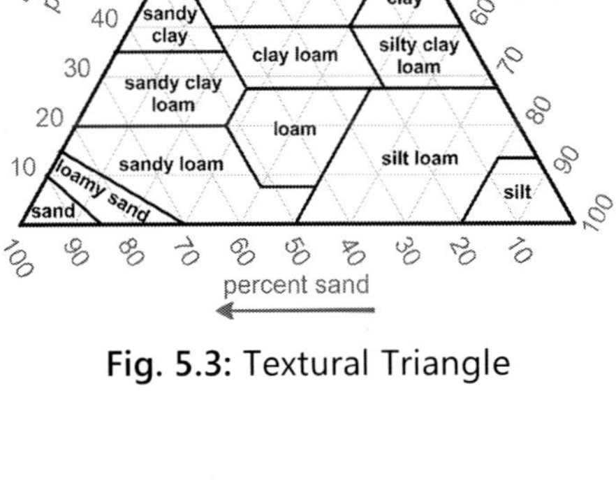

Fig. 5.3: Textural Triangle

5.3 Soil Moisture and its retention in soil

Presence of moisture does not indicate its availability to the plant. The water in the soil is held by different forces that impacts its retention and movement. The water molecule consists of hydrogen and oxygen ion. The oxygen ion surrounds the hydrogen ion and acts as an attracting force. The forces that holds the water with soil particles can be categorized as cohesive and adhesive forces. Because of these forces there can be three different types of water. Fig. 5.4 shows different types of soil water.

5.3.1 Kinds of water

(a) Hygroscopic water: Water that is tightly held with the soil particles due to force of adsorption are known as hygroscopic water. This water cannot be used by the plants.

(b) Capillary water: The capillary water is not as tightly held with the soil particles. The major forces acting are surface tension and they are available in continuous films around the soil particles in the capillary pores. This water is available for use by the plant

(c) Gravitational water: As the thickness of water layer around the soil particles increase the gravity forces dominates and it starts flowing freely through the soil matrix.

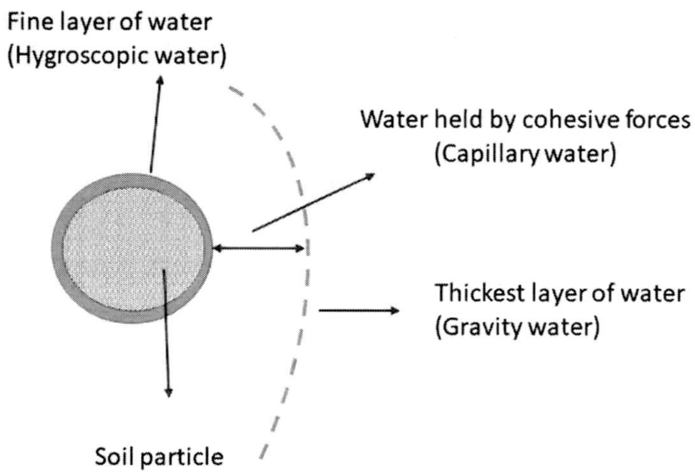

Fig. 5.4: Diagrammatic representation of different soil water

When the moisture content is high the dominating force is surface tension and in dry range the force of adsorption is dominating. The water retained in the soil and move within it is the result of energy effect.

5.3.2 Surface tension

Surface tension can be defined as the tractional force, γ, acting across any unit length of line on this fictitious membrane (Ono and Kondo 1960). Surface tension is a force pulling inward at surface of a liquid, tending to make the surface area as mall as possible. Surface tension is the energy per unit area and is measured in ergs/cm^2.

If P_a and P_b are the forces acting on the inside and outside of a water bubble, the difference in force (P_a-P_b) acting on the entire area of the cross section of the bubble (πr^2) is the force resisting the inward pull due to surface tension (σ). Equating the two forces

$$4\pi r\sigma = (P_b - P_a)\ \pi r^2 \qquad \text{...... (5.1)}$$

$$P_b - P_a = \frac{4\sigma}{r} \qquad \text{...... (5.2)}$$

As the radius increases the internal pressure also increases within a given temperature limits. That is, the surface tension force is balanced by a pressure drop across the air-water interface.

5.3.3 Viscosity

Viscosity is the internal resistance offered to flowing water. The pressure forces of flowing water are counteracted by the drag forces. Viscosity is function of temperature.

5.3.4 Capillarity

The capillary water is the water retained in soil after gravitational water is drained out. The water remains are subject to law of capillarity by forming layers around soil grain. Capillarity is a result of phenomenon of surface tension. The capillary water is due to the attraction of water in hair like openings or capillary pores. Capillarity depends on cohesive and adhesive forces. The capillarity explains the height of rise of water in soil pores.

If we consider a capillary tube, the water in the tube is held by the forces of surface tension and resisting the gravitational pull due to weight of the water. The vertical component of surface tension force is $2\pi r\sigma\cos\theta$. The weight of the column on water of height h and diameter 2r can be given as $g\pi\,\rho r^2 h$, where ρ is the density of water and g is the acceleration due to gravity. Equating the two forces

$$2\pi r\sigma\cos\theta = g\pi\,\rho r^2 h \qquad \text{...... (5.3)}$$

$$\text{h} = \frac{2\sigma\cos\theta}{g\rho r} \qquad \text{...... (5.4)}$$

The equation helps to understand the rise of water in a capillary tube. The diagrammatic representation is shown in Fig. 5.5 and Fig. 5.6.

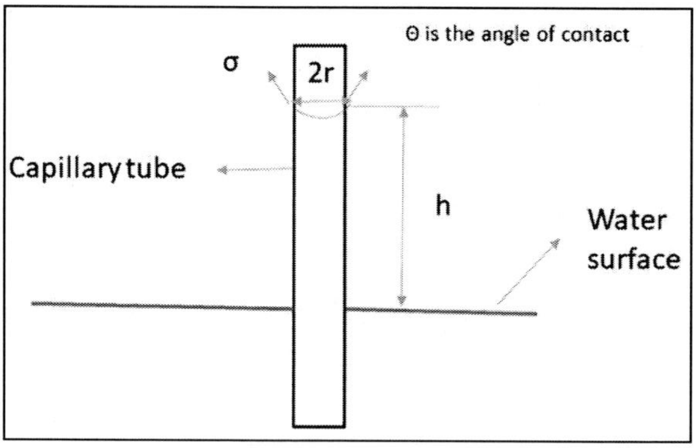

Fig. 5.5: Rise of water in capillary tube

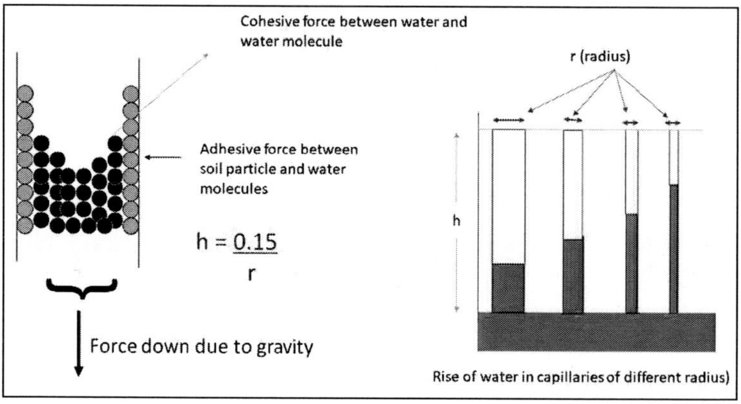

Fig. 5.6: Capillary rise

5.4 Energy concept of soil water

The water present in the soil is subjected to different forces. Since the rate of flow of water within the soil is slow, the kinetic energy potential is considered negligible and thus only potential energy is considered. Therefore, total energy is considered as equivalent potential energy. The equivalent potential is sum of different energy is due to virtue of its position in the field.

5.4.1 Definition of Total soil water potential

Total soil water potential is defined as the amount of work per unit quantity of pure water that must be done by external forces to transfer reversibly and isothermally an infinitesimal amount of water from the standard state to the soil at the point under consideration. Fig. 5.7 shows the different potential and its position with respect to reference level.

The water present in the soil is being acted upon by different forces at different points and hence the potential energy is not constant but variable in nature. Since the work has to be done for moving water from a reference elevation to a higher elevation within the soil medium, the potential may be positive or negative.

The different potential energy active in a soil matrix are:

(i) Gravity potential (ψ_g)

(ii) Pressure potential (ψ_p)

(iii) Capillary potential/ matric potential(ψ_m)

(iv) Osmotic potential (ψ_o)

The total potential ψ_T can be expressed as $\varphi_T = \varphi_g + \varphi_m + \varphi_p + \varphi_o$

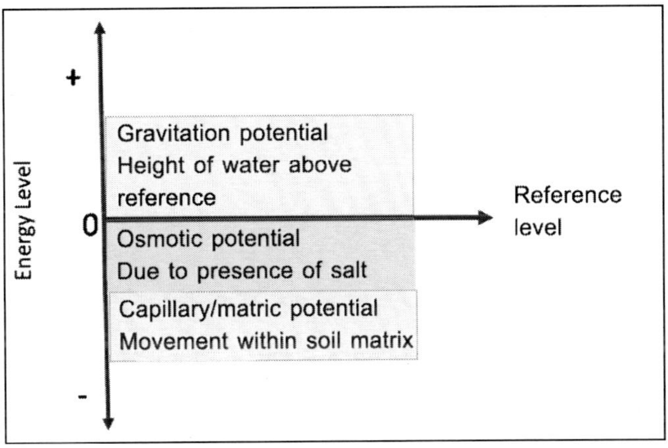

Fig. 5.7: Different potentials and its position with respect to reference

5.4.1.1 Gravity Potential:

The gravitational potential is attributed to the gravitational forces acting on the water molecule and is dependent on the elevation with respect to the equilibrium condition (commonly the ground water level). The gravitational potential can be defined as:

$$Potential(E_g) = Force \times Distance = mgl = \rho_w Vgl \qquad ......(5.5)$$

Where, ρ_w is the density of water in g/cm^3; V is the volume of water in cm^3; g is the acceleration due to gravity cm/sec^2 and l is the height of water above reference level.

The potential can be expressed in either per unit mass, per unit volume or per unit weight.

$$\varphi_g\,(mass) \text{ per unit mass} = \frac{E_g}{M} = \frac{mgl}{m} = gl \qquad \text{...... (5.6)}$$

$$\varphi_g\,(volume) \text{ per unit volume} = \frac{E_g}{v} = \frac{\rho_w Vgl}{v} = \rho_w gl \qquad \text{...... (5.7)}$$

$$\varphi_g\,(weight) \text{ per unit volume} = \frac{E_g}{W} = \frac{\rho_w Vgl}{\rho_w Vg} = l \qquad \text{...... (5.8)}$$

The diagrammatic representation of gravitational potential is shown in Fig. 5.8.

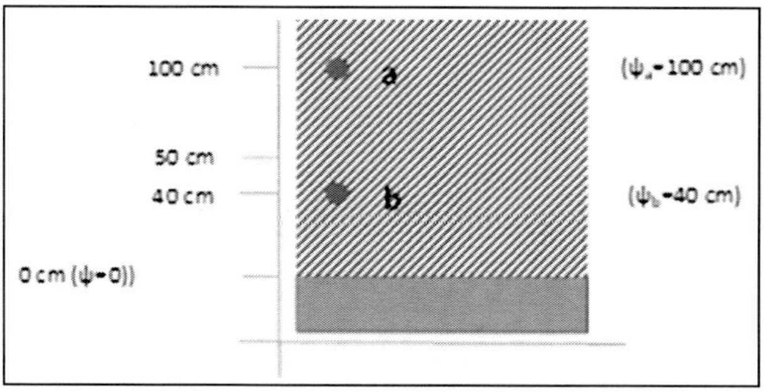

Fig. 5.8: Gravitational potential is the position of water with respect to reference level

5.4.1.2 Pressure/Capillary/ matric potential:

Capillary or matric potential can be defined as Energy per unit volume of water required to transfer an infinitesimal quantity of water from a reference pool of water at the elevation of the soil to the point of interest in the soil at reference air pressure and temperature. The capillary potential can be positive or negative. If the soil is saturated the potential is known as pressure or submergence potential and if it is unsaturated the potential can be expressed as matric or capillary potential. Fig. 5.9 shows the capillary rise in a soil.

The pressure potential when the soil is saturated is known as hydrostatic or pressure potential. When the soil is unsaturated the potential acting is known as capillary potential. It is negative in nature. This potential is attributed to soil matrix.

The hydrostatic pressure with respect to atmospheric pressure can be expressed as:

$$E_p = \rho_w gh \qquad \text{...... (5.9)}$$

The pressure potential expressed in mass, volume and weight unit is as follows:

$$\varphi_p(mass) \text{ per unit mass} = \frac{E_p}{M} = \frac{\rho_w gh}{\rho_w} = gh \qquad \text{...... (5.10)}$$

$$\varphi_p(volume) \text{ per unit mass} = \frac{E_p}{v} = \frac{\rho_w ghd_v}{v} = \rho_w gh \qquad \text{...... (5.11)}$$

$$\varphi_g(weight) \text{ per unit volume} = \frac{E_p}{W} = \frac{\rho_w gh}{\rho_w g} = h \qquad \text{...... (5.12)}$$

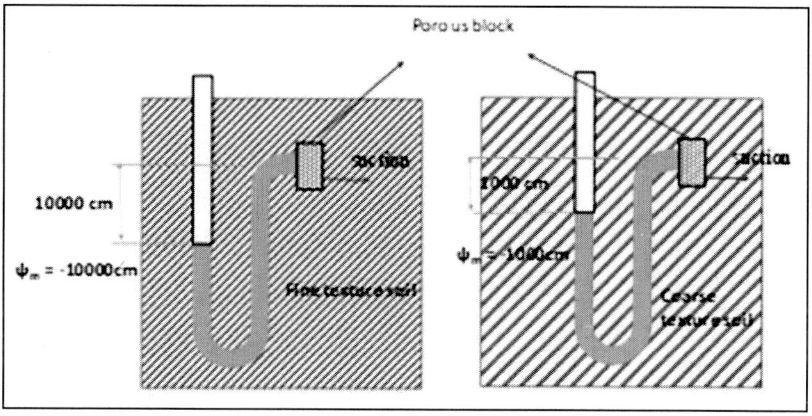

Fig. 5.9: Matric potential is the vertical distance between the water surface and porous cup

5.4.1.3 Osmotic potential:

Osmotic potential can be defined as measure of potential to move water within different concentration across a water permeable membrane. Osmotic potential results from the presence of salts in the soil. The salts present in the soil alter the free energy. Thus with the change in extent of free energy the movement of the water reverses from plant to soil. It is also known as solute potential and is negative pressure. It can be expressed mathematically as:

$$\varphi_O = -CRT = -\pi \qquad \text{...... (5.13)}$$

Where, C is the concentration of solute, R is the universal gas constant (i.e. 8.314472 J K^{-1} mol^{-1}), and T is the absolute temperature.

5.5 Soil Moisture Tension

Water has a tendency to move from state of high free energy to low free energy. Thus the potential in the soil decides the movement of water from soil to plant or plant to soil. The water present in the soil is subjected to different forces and the plant has to exerts force to extract water from the soil for its metabolic activities. It is expressed as soil moisture tension or stress.

Fig. 5.10 shows the movement of water from higher potential to lower potential.

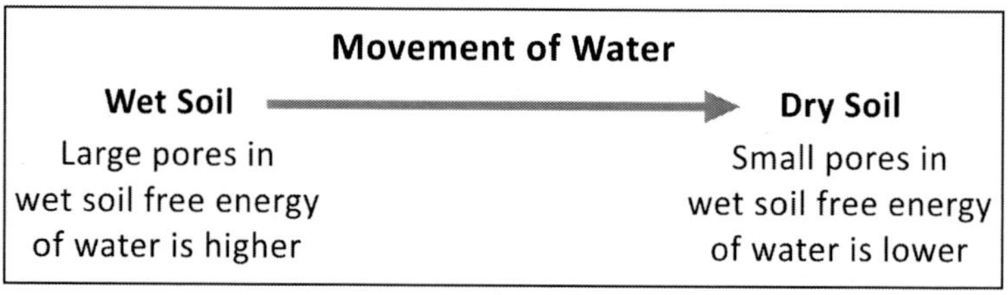

Fig. 5.10: Movement of soil water from higher potential to lower potential

When the force required increases the plants finds difficulty in extracting water from the soil. Thus the drooping of leaves starts and it is generally termed as wilting. Under wilting conditions, the metabolic activities of the plants are restricted that may impact the productivity. Generally, a plant is comfortable when the potential is within 1 atm. However, as the potential increases the plant find difficulty in extracting water and we term it as soil moisture stress. The potential is not constant but vary with the soil moisture content. The extent of moisture contents is described as soil water constants.

5.6 Soil Moisture Constants

Different soil moisture constants are described below

(i) ***Saturation Capacity:*** Just after irrigation or heavy rainfall all the soil pores are filled with water. Under this situation gravitation potential are dominant and water starts moving downwards. At saturation capacity, the total potential is almost zero. The situation is known as saturation capacity.

(ii) ***Field Capacity (FC):*** The pores present in the soils are of different sizes. The water present in larger pores are held loosely as compared to micro pores. After the process of irrigation or rain stops the water present in macro pores moves downward due to force of gravity. Such situation arises after 24 to 48 hours of saturation capacity. Under this situation, the macro pores are filled up with air and micro pores are filled with water. The force with which the water is held is in between 0.1 to 0.3 atmosphere (atm). This condition is known as *field capacity*. At field capacity the plant can extract water from soil very easily.

(iii) ***Permanent wilting point (PWP):*** The vegetation present in the soil starts consuming soil moisture present at field capacity. Slowly the moisture content decreases and the soils starts becoming unsaturated. At unsaturated conditions, the capillary forces start

becoming dominant. As the soils starts becoming dry the moisture is tightly held by the soil particles thereby increasing the total potential. As and when the potential exceeds 7 atm the plants finds difficulty in extracting soil moisture and thus restricts its metabolic activities. Such situations are known as *permanent wilting point.*

The PWP varies from plant to plant. Some plants start wilting at 7 atm and some plants wilt at 15 atm. Even shallow rooted crops like vegetables may start wilting at as low as 5 atm soil moisture potential.

Apart from the reduction in soil moisture content the extent of free energy may change depending on the presence of salts. As the salt concentration increases the free energy within the plant body remains more than the free energy of water in soil. Under such conditions the plant is unable to extract water from the soil. At PWP if water is provided to the soil the plant may recover from the condition of wilting

(iv) ***Ultimate Wilting Point (UWP):*** As the soil potential exceeds 15 atm and water is not added to the soil, the hygroscopic forces that holds the water tightly with soil particles increases. At this stage the plant is unable to extract water from the soil. Generally, the UWP ranges from 15 to 32 atm or at times it may go as high as 60 atm. At UWP is the plant is unable to recover and dies.

(v) ***Available Water Capacity (AWC):*** Soil moisture content between Field Capacity (FC) and Permanent Wilting Point (PWP) is known as Available Water Capacity (AWC). The AWC is the moisture content that is available in the soil for use by a plant for its metabolic activities. The AWC varies from soil to soil, in sandy soil the AWC is less as compared to clay soil.

(vi) ***Moisture Equivalent:*** As per American Meteorological Society moisture equivalent is the weight percentage of water (with respect to dry weight) retained by an initially saturated 1-cm-thick soil sample after it has been subjected to a centrifugal force 1000 times the force of gravity for 30 minutes. It can be determined by exposing a saturated soil sample in centrifuge with a centrifugal force 1000 times the force of gravity for 30 minutes.

As the soil is subjected to centrifugal force it releases water till increase in capillary forces are balanced by centrifugal forces acting on soil moisture. Under this condition a capillary equilibrium is achieved. The moisture content at this condition is known as moisture equivalent (Briggs and McLane, 1910). The moisture equivalent varies from soil to soil. In loamy soil the moisture equivalent is equal to field capacity (FC), where as in sandy soil it is higher and clay soil it is lower than field capacity (FC). Fig. 5.11 shows the stages of water holding.

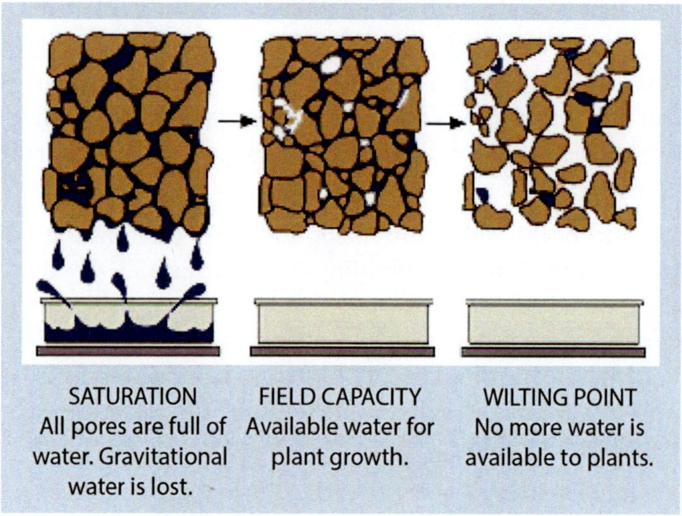

Fig. 5.11: Stages of water holding
(adapted from Vengadaramana and Jashothan, 2012)

5.7 Soil Moisture Characteristics Curve

Soil moisture characteristics curve (or retention curve) is relationship between the soil water potential and volumetric water content. The soil moisture characteristics curve is determined by measuring the moisture content at different potential and is plotted. The curve is function of different properties of soil such as texture, structure and other characteristics of soil. When a saturated soil sample is subjected to external pressure the large pores drains first. If the pressure is continuously increased the diameter of the water filled pores decreases. The water contained in these pores are held by capillary forces. At zero potential the soil water content is defined as saturation constant and as the potential increases the soil starts getting unsaturated and moves towards permanent wilting point. A typical soil moisture characteristics curve is shown in Fig. 5.12.

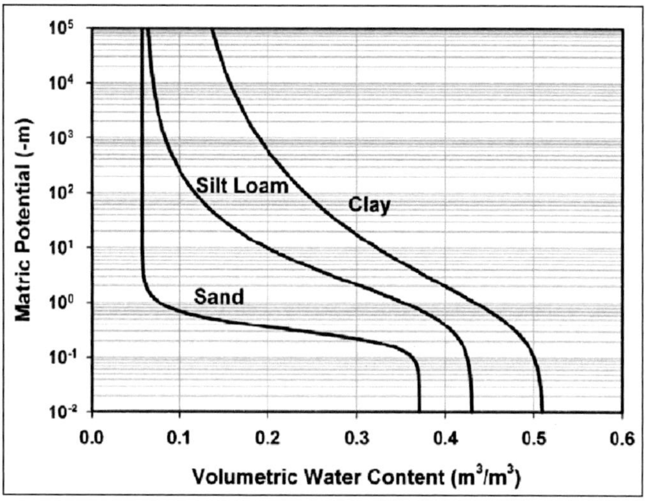

Fig. 5.12: Typical soil water characteristic curves for soils of different texture (adapted from Tuller *et al.*, (2004)

5.8 Measurement of Soil Moisture retention

5.8.1 Pressure plate pressure membrane apparatus

Pressure plate pressure membrane apparatus (Fig. 5.13) is used to estimate soil moisture retention of entire pressure range from 0 to 15 atm. It consists of an air tight chamber with a ceramic pressure plate or membrane. Saturated soil sample is placed on the ceramic plate and known pressure is applied externally with the help of a compressor. Fig. 5.14 shows the schematic diagram of pressure plate apparatus. After application of known pressure to the sample the moisture content is determined by gravimetric method. This apparatus enables to measure soil moisture for entire range of pressure. The results obtained can be used to develop soil moisture characteristics curve.

Fig. 5.13: Pressure plate pressure membrane apparatus

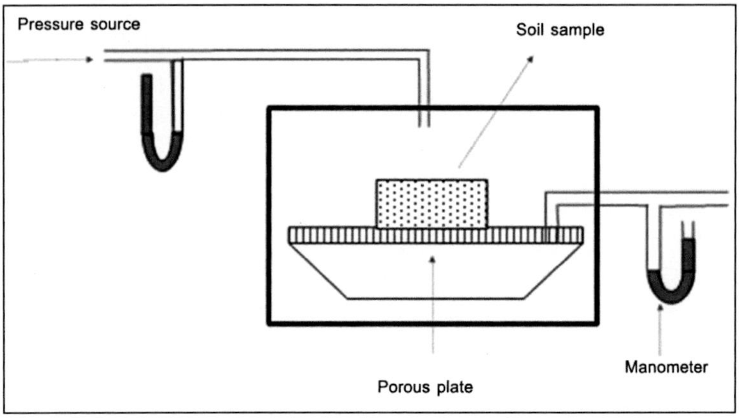

Fig. 5.14: Schematic diagram of pressure plate apparatus

5.8.2 Tensiometers

Tensiometers are simple device that measures the direct tenacity (tension) of the soil. It measures the capillary/ matric potential of the soil.

It consists of an access tube of desired length. At the end of tube, a porous cup is attached. The tube is closed with an air tight cap. The tube is attached to a vacuum/pressure gauge (Fig. 5.15). The tube is filled with water and is placed in the soil at desired depth. When the porous cup comes in touch with the dry soil, the water starts moving out till an equilibrium condition arises between the cup and soil. As the water comes out of the tube a vacuum is created within it and it is reflected in the vacuum gauge attached to it.

A tensiometer can measure the tensions in the range of 0 to 0.85 atmosphere. A reading of near to 0 indicate wet (saturated) soil whereas 0.85 indicate dry conditions. Beyond 0.85 a tensiometer becomes nonfunctional due to discontinuation of water column within the tube.

Advantages

(a) Direct reading is possible
(b) The observation can be automated
(c) High skills not required
(d) Not affected by presence of salts in the soil
(e) High frequency sampling is possible
(f) Low cost.

Disadvantages

(a) Measurement limit less than 1.0 atm (100 kPa)

(b) Relatively slow response to soil moisture potential in the soil

(c) The porous cup should be in touch with the soil for obtaining reading

(d) Requires frequent filling of the tube.

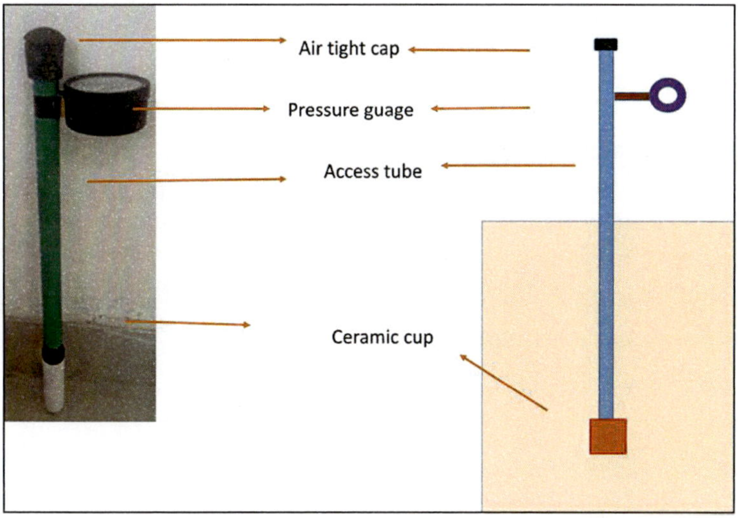

Fig. 5.15: Tensiometers and its component

5.8.3 Soil moisture sensors (*Time Domain reflectometry, TDR*)

TDR method for the measurement of soil water content becomes more and more popular for its simplicity of operation, accuracy and non-destructiveness, as compared to other methods. The TDR (Topp et al., 1980) measures the propagation of an electromagnetic wave through the soil. The characteristics of this propagation depend on soil water content through the dielectric properties of the soil. Advantage is it gives real time moisture measurement. The relation between the soil moisture and its dielectric constant is highly correlated for the majority of soils. It can also measure a wide range of soil moisture content (Skierucha, 2000). It gives the volumetric moisture content of soil (Fig. 5.16).

Advantages

(a) They do not require maintenance as they are placed below ground

(b) Response to change in soil moisture is fast

(c) It can measure soil temperature and soil electrical conductivity (EC).

Limitations

(a) The error in the measurement of volumetric water content by the TDR technique results from the correlation imperfections between the directly measured values of soil refractive index, n, and the real value of soil moisture as well as the hardware and software imperfections of the TDR device and TDR probe installation. Inclusion of the soil bulk density in the TDR calibration formula decreases the absolute error of the TDR determined soil moisture by the factor of two.

(b) Readings are affected by presence of salts. In soils with EC values more than 1.32 dS/m results in errors in reading (McIsaac 2010).

(c) The relative error of TDR moisture values increases in the lower range of water contents. This is due to a constant absolute error introduced by the measuring device and an increasing role of the soil solid phase in the soil refractive index.

(d) Air gaps if any between sensor and soil during installation may give erroneous readings.

(e) It many a times is beyond affordable range of farmers.

Fig. 5.16: Soil moisture sensors and data logger

5.8.4 Neutron scattering

Neutron probes use the property of scattering and slowing down neutrons by the hydrogen nuclei of the water molecules. Its main advantages are that non-destructive and direct measurements can be performed without disturbing the soil. This allows one to follow water content changes with time. This equipment can be considered affordable since it can be used at several locations in one field and for several fields, once access tubes are installed and conveniently distributed in the field.

The main limitation relates to safety rules which have to be followed to operate, transport and store the probe. This is the main reason why there is a tendency to replace this equipment by the recently developed Time Domain Reflectometry apparatus (TDR).

Example: Two tensiometer is placed at two depth viz., 100 cm and 40 cm from the reference level (Fig. 16). The reading for the matric potential obtained at the two location are -115 cm and -30 cm respectively. What is the total potential and in which direction water will move?

As the depth of two points a and b are 100 cm and 40 cm the gravitational potential ψ_{ga}= 100 cm and ψ_{gb}= 40 cm.

The matric potential ψ_{ma}= -115 cm and ψ_{mb}= -30 cm respectively

Total potential ψ_{Ta}= 100+(-115)= -15 cm and ψ_{Tb}= 40+(-30)= 10 cm

Total potential ψ=ψ_{Ta}+ ψ_{Tb}= -15+10= -5 cm

So water will move upward (Fig. 5.17).

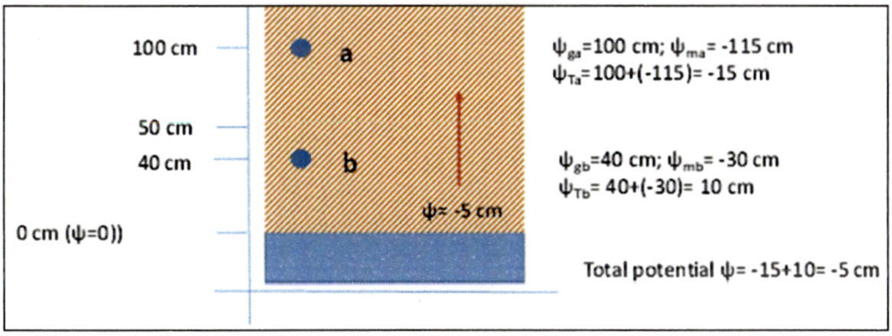

Fig. 5.17: Matric potential and direction of flow of water due to capillarity in soil matrix

Chapter - 6

THERMODYNAMICS

6.1 Basics of Thermodynamics

Thermodynamics can be defined as the study of energy, energy transformations and its relation to matter. In other words, is the study of the effects of work, heat, and energy on a system.

All the thermodynamics systems can be expressed in four quantities, *viz.* temperature (T), internal energy (U), entropy (S) and heat (Q). The classical thermodynamics concern with the relationships between the bulk properties of matters, it doesn't concern with any properties at microscopic level. Thermodynamics only provides macroscopic descriptions of the states of complex systems and of their behaviours when they interact or are constrained under various circumstances. A thermodynamic system is a precisely specified macroscopic region of the Universe. It is limited by boundaries of a particular nature with particular properties. All space in the Universe outside the thermodynamic system is known as the surroundings, the environment, or the reservoir. A thermodynamically isolated system is completely isolated from its environment: it does not exchange energy or matter with the environment. If matter is able to flow in or out of its boundaries, the system is said to be open. Otherwise it is said to be closed. Processes that affect the interior of the region of the system are studied using general principles: the four laws of thermodynamics. Although historically established before modern physics, most of these laws are based on the fact that a system is governed by general statistical principles which apply to the random motion of its particles. Hence, two approaches are possible when studying a system by consideration of its thermodynamics: a very general and practical one is to establish balances of the transfers between the system and its surroundings; the other is to apply statistical laws to the system. The latter approach is suitable for simple systems, and may also be used to clear up some of the more abstract concepts within thermodynamics.

Before discussing soil as a thermodynamic system, let us explain certain definitions associated with thermodynamic systems.

System: A system is defined as a quantity of matter or a region in space chosen for study. The mass or region outside the system is called the surroundings. Fig. 6.1 shows the system, boundary and surroundings.

Boundary: The real or imaginary surface that separates the system from its surroundings. The boundaries of a system can be fixed or movable. Mathematically, the boundary has zero thickness, no mass, and no volume.

Closed system or control mass: It consists of a fixed amount of mass, and no mass can cross its boundary. But, energy in the form of heat or work, can cross the boundary, and the volume of a closed system does not have to be fixed.

Open system or control volume: It is a properly selected region in space. It usually encloses a device that involves mass flow. Both mass and energy can cross the boundary of a control volume.

Isolated system: A closed system that does not communicate with the surroundings by any means.

Rigid system: A closed system that communicates with the surroundings by heat only.

Adiabatic system: A closed or open system that does not exchange energy with the surroundings by heat. Closed system, mass cannot cross the boundaries, but energy can.

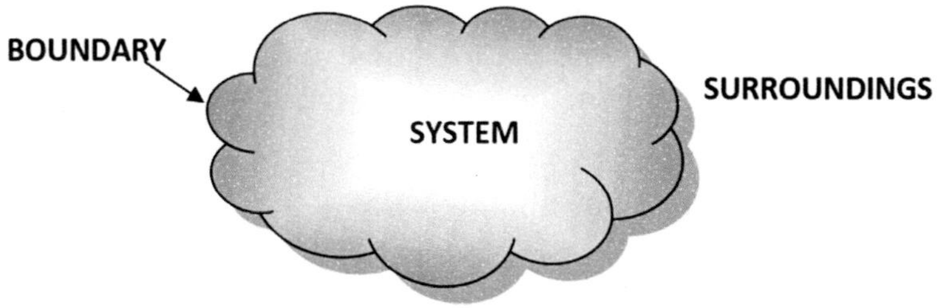

Fig. 6.1: System, boundary and surroundings

A thermodynamic system is governed by certain set of laws called the laws of thermodynamics. These laws explain the behaviour of the system with the surroundings based on its state. First law primarily explains the conservation of energy or one can say that it is an extension of the Law of Energy Conservation. The first law states that the **change in internal energy of a system is equal to the heat added to the system minus the work done by the system, i.e.**

$$\Delta U = Q - W \qquad \text{...... (6.1)}$$

In other words it simply states that during an interaction, energy can, change from one form to another but the total amount of energy remains constant. Any thermodynamic system in an equilibrium state possesses a state variable called internal energy (U). Between any two equilibrium states, the change in internal energy is equal to the heat transfer into the system and work done by the system. In any thermodynamic system there will be one of the following four processes. It may be adiabatic where there is no heat transferred between the system and the surroundings, i.e. the system is insulated from the surroundings, and a process is called isothermal if it occurs at constant temperatures. Isobaric process of thermodynamic system occurs at constant pressure and the isochoric process occur at constant volume.

An adiabatic process transfer no heat and therefore loss of heat is zero. When a system expands adiabatically, the work done (W) is positive and so the change of internal energy becomes negative as the system does the work in adiabatic system. On the other hand when the system compresses adiabatically the change of internal energy becomes positive as the work done within the system becomes negative. An isothermal process is a constant temperature process. Any heat flow into or out of the system must be slow enough to maintain thermal equilibrium. For ideal gas, if ΔT is zero then the change in internal energy (ΔU) will also be zero and hence any heat entering into the system must convert into work done. In isobaric process which occurs at constant pressure, the change in internal energy, heat entering into the system and the work done are nonzero quantities. For ideal gas the work done results is change of volume at constant pressure. In isochoric process as the volume doesn't change there will not be any work done to the surrounding and hence the change of internal energy is equal to the heat exchange into or out of the system with the surrounding.

Energy has quality as well as quantity, and actual processes occur in the direction of decreasing quality of energy. The second law of thermodynamics says that there exists a useful thermodynamic variable called entropy (S). A natural process that starts in one equilibrium state and ends in another will go in the direction that causes the entropy of the system plus the environment to increase for an irreversible process and to remain constant for a reversible process.

$$S_f = S_i \text{ (reversible) and } S_f > S_i \text{ (irreversible)}$$

Where S_i is the initial entropy and S_f is the final entropy of the system undergoing the process.

Mathematically, the second law is expressed as:

$$\Delta S = \frac{\Delta Q}{T} \qquad \text{...... (6.2)}$$

The second law uses the notion called entropy (S) which is the disorder or randomness of the system. The second law helps to determine the direction of the processes. A reversible process is those which can change the state and comes back to the original state.

According to Carnot Cycle, the process from 1 to 2 is isothermal expansion where work is done by the gas (Fig. 6.2). The process from 2-3 is adiabatic expansion where work is again done by the gas. In the process 3-4, the isothermal compression takes place and work is done on the gas. In the process from 4-1, it is adiabatic compression and work is done on gas.

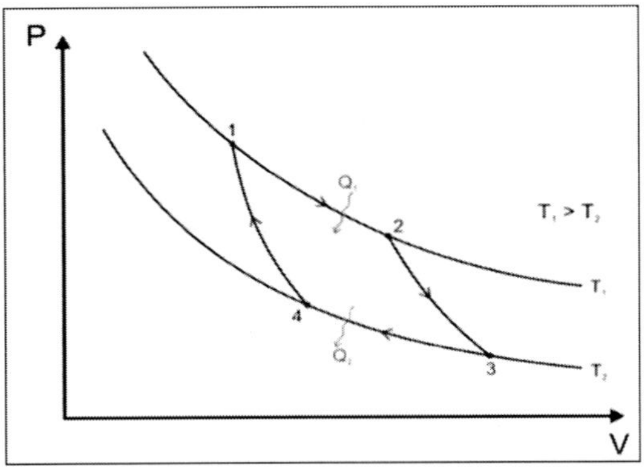

Fig. 6.2: The Carnot Cycle

The Carnot cycle is a theoretical thermodynamic cycle proposed by French physicist Sadi Carnot in 1824 and expanded upon by others in the 1830s and 1840s. It provides an upper limit on the efficiency that any classical thermodynamic engine can achieve during the conversion of heat into work, or conversely.

The third law of thermodynamics states that it is not possible for any system to reach absolute zero. The third law is also expressed in terms of entropy. It says that the entropy of all the system is the same at absolute zero. The law also explains that internal energy of any system is proportional to the temperature.

Another law of thermodynamics was derived when the first and second laws were already established. Through the new law specifically explained the basic principle of thermodynamics i.e. equilibrium of states, the new law could not be named as first or second law. The zeroth law of thermodynamics states that if objects A and B are in thermal equilibrium with another object C, then the objects A and C are in thermal equilibrium with each other. This law allows us to define temperature relative to an established standard.

6.2 Soil as a thermodynamic System

Soil is a three phase system which is the assembly of solid, liquid and gases. This is also true from the thermodynamic view point. Further it also consists of the fields of electromagnetic and gravitational energy. These characteristics, together with a surface that encloses the macroscopic region of space filled by the soil, define the thermodynamic soil system. Thus a thermodynamic soil system contains both matter and physical fields and is bounded by a surface of arbitrary shape. This bounding surface is called the thermodynamic wall surrounding the soil system. The properties of a thermodynamic soil system by definition refer to a macroscopic region of space and do not depend for their measurement or interpretation on the previous history of the system. It is important to understand that the thermodynamic properties constitute only a part of all the properties of a soil. Those soil properties that describe phenomena at the molecular scale or those that relate directly to the effect of time are not included. Therefore the concentration of protons and the electric potential at a point of few nanometres away from the surface of a soil colloid suspended in water are not properties of a thermodynamic soil system.

The properties of soil system may be divided into two types, one is fundamental properties and other is derived properties. This division is however arbitrary. The fundamental properties of a soil always have the distinguishing feature that together they make up the smallest set of properties that provide a complete thermodynamic description but yet can be varied independently. A set of numerical values of the fundamental properties are known as the state of the system. A system is also defined by the parameters or properties. The properties whether fundamental or derived depend on the matter or mass in the system are called extensive properties (e.g. volume, entropy) and the properties not dependent on the mass in the system are called intensive properties (e.g. pressure, temperature). The intensive properties of a soil are mathematical field variables (i.e., their values are associated with points in space that are located in the soil). Since a thermodynamic property always refers to a macroscopic region of space, it is understood that the value of an intensive property at some point in a soil applies to a macroscopically small neighbourhood of that point that encloses many solid grains and interstices. If the values of its intensive properties are the same everywhere in a soil, the soil is said to be homogeneous. If anyone of its intensive properties varies (on the macroscopic scale) from point to point, the soil is said to be heterogeneous.

Soil as a natural body is invariably heterogeneous. The intensive properties of the soil system vary spatially at macroscopic scale due to pedo-chemical processes or due to direct effect of gravitational field of the earth. However, thermodynamic soil system is considered as a homogeneous system for all practical purposes. Soil consists of components. These are materials of fixed chemical compositions and properties which usually vary in soils. The soil compositions are therefore specified in terms of its components. A homogeneous portion of a soil that has a variable composition is called a solution. Thus the gaseous, liquid, and solid portions of a soil each may be solutions. The interstices in the soil may contain air, a gaseous solution composed principally of nitrogen, oxygen, carbon dioxide, and water vapor, as well as the soil solution, an aqueous solution composed of liquid water and dissolved solids. The solid portion of a soil also may be a solution if its composition is mixed.

The thermodynamic wall surrounding a soil is a very important part of the system. If the wall permits the free transfer of both matter and thermal energy either in or out, the soil is called an open system and the wall is said to be diathermal and permeable. If only certain types of matter may be transferred through the wall, it is said to be semipermeable. If the wall permits only the transfer of thermal energy, it is said to be diathermal and the soil is called a closed system. Finally, if the wall does not permit the transfer of either matter or thermal energy, it is said to be insulating and the soil is called an adiabatic system. Note that mechanical energy or the energy associated with gravitational and electromagnetic fields may be transferred through any of the walls that have been discussed. Thermodynamic walls are differentiated only by their behavior toward the transfer of matter and thermal energy.

6.3 Thermodynamic Processes in Soil

When thermodynamic properties of soils are changed in some ways, then a thermodynamic process takes place. Thermodynamic processes definitely result in certain change of state in the system. If the process passes exclusively through the state of equilibrium, the thermodynamic process is called reversible process and such processes are usually limiting cases of any natural system. As explained in Section 4.1, the thermodynamic processes in soil system also have adiabatic, isothermal, isobaric and isochoric processes.

Soil thermodynamic phenomenon can be explained with the help of a set of soil properties these are:

Temperature: In the event of energy transfer in the thermodynamic phenomenon, soil temperature becomes a criterion of equilibrium. It is an intensive property of soil and expressed in Kelvin (K).

Entropy: Entropy is an extensive property. It is the principal criterion of thermodynamic equilibrium. It is also defined as the randomness of the molecules in a system. Entropy is expressed in terms of Joule per Kelvin.

Pressure: Pressure is the criterion of mechanical equilibrium. Pressure is an intensive property of a system and expressed in terms of Newton per square meter.

Volume: Volume is an extensive property of a system. It expresses the spatial extent of soil and therefore forms the basis of assessing mechanical energy transfer to and from the soil. It is expressed in terms of m^3 or cm^3 or litre.

Chemical Potential: It is an intensive property. The chemical potential of soil sets the criterion of equilibrium with respect to transfer of energy and matter. Every component of soil carries chemical potential which determines the way the component takes or transforms its phases from one another or forms a entirely new compound. Similar to the thermal energy transfer based on its potential, the chemical potential also gets transferred from higher potential to lower potential phases. Chemical potential usually measured in terms of Joules per mole.

Mass: Mass is an extensive property of soil system. Mass is expressed in terms of kilograms (kg). Often the soil components are also expressed in terms of mole number. But unless the chemical reactions are considered, the mole number may not be a useful unit in thermodynamic processes.

These six intensive and extensive properties are considered to be primary in thermodynamic analysis. Other properties are actually derived properties. For example bulk density is an important property at macroscopic scale but it is a derived property from the volume and the mass. The thermodynamic properties of soil are given in the tabular form (Table 6.1)

Table 6.1: Thermodynamic properties of soil

Variable	Symbol	Extensive	Intensive	Unit
Temperature	T		X	K
Entropy	S	X		JK^{-1}
Pressure	P		X	Nm^{-2}
Volume	V	X		m^3
Chemical Potentials	μ		X	Jkg^{-1}
Mass	m	X		kg

6.4 Thermodynamic Potentials

The thermodynamic concept of internal energy and entropy as applied to the other standard thermodynamic system are also applied to soil system without any modification. As per the first and second laws of thermodynamics, the internal energy of a system is a function of entropy, volume and compositions and is given by:

$$U = f(S, V, \{m_{i,\alpha}\}) \qquad \text{...... (6.3)}$$

$\{m_{i,\alpha}\}$ indicates all the components mass wherein i is component index and α is phase index of the component.

The Eq. 4.1 is a fundamental relation for a soil because the equilibrium states of the system are characterized completely once the dependence of internal energy (U) has been established on the independent variables of state S, *V,* and $\{m_{i,\alpha}\}$. Fundamental relations often may be developed by investigating infinitesimal processes, whose complete description is contained in the differential form:

$$dU = TdS - PdV + \Sigma_i \Sigma_\alpha \mu_{i,\alpha} dm_{i,\alpha} \quad \text{...... (6.4)}$$

Herein the assumption is that all the phases are in same temperature and pressure . If the assumption is not true than the phases are to be considered and accordingly the first term in the right hand side will be replaced with summation of all phase entropies in differential forms. The coefficients in the eq. 6.4, T, P and $\mu_{i,\alpha}$ can be expressed as the partial derivatives of the internal energy of the system or they can be expressed as the functions of all independent variables. i.e

$$T = T(S,V,\{m_{i,\alpha}\}),\quad P = P(S,V,\{m_{i,\alpha}\}),$$

$$\mu_{i,\alpha} = \mu_{i,\alpha}(S,V,\{m_{i,\alpha}\}) \quad \text{...... (6.5)}$$

These equations 6.5 are called equations of state.

When the equation 6.3 is operated with Partial Legendre Transformation, a set of systematic equations can be formed which are called "Thermodynamic Potentials".

The characteristics of Thermodynamic potentials are given as:

1. The thermodynamic potentials are extensive property and equivalent of internal energy of the system. Complete thermodynamic properties of soils can be explained through thermodynamic potentials.
2. Each of the thermodynamic potentials is uniquely determined by independent variables of state which may be chosen to explain the soil system.
3. The thermodynamic potentials have the property that the equilibrium state of a soil, in contact with reservoir that controls the intensive variables on which potential depends. Any decrease of thermodynamic potential is equal to the reversible works in the soil system.

The three important thermodynamic potentials obtained through partial Legendre Transformation of Eq. 6.3 are Enthalpy, Helmholtz Energy and Gibbs Energy.

Enthalpy: With S, P and $\{m_{i,\alpha}\}$ as independent variables of state, the resulting thermodynamic potential is called Enthalpy. Enthalpy is expressed as:

$$dH = TdS + VdP + \Sigma_i \Sigma_\alpha \mu_{i,\alpha} dm_{i,\alpha} \qquad \text{...... (6.6)}$$

and the corresponding state equations are:

$$T = T(S, P, \{m_{i,\alpha}\}), \quad V = V(S, P, \{m_{i,\alpha}\}),$$

$$\mu_{i,\alpha} = \mu_{i,\alpha}(S, P, \{m_{i,\alpha}\}) \qquad \text{...... (6.7)}$$

A decrease of enthalpy in a soil contacting a volume reservoir that controls the applied pressure, *P*, is equal to the reversible work delivered by the soil.

Helmholtz Energy: With T, V and $\{m_{i,\alpha}\}$ as independent variables of state, the thermodynamic potentials is called as Helmholtz Energy (A). This is expressed as:

$$dA = -SdT + PdV + \Sigma_i \Sigma_\alpha \mu_{i,\alpha} dm_{i,\alpha} \qquad \text{...... (6.8)}$$

and the corresponding state equations are:

$$S = S(T, V, \{m_{i,\alpha}\}), \quad P = P(T, V, \{m_{i,\alpha}\}),$$

$$\mu_{i,\alpha} = \mu_{i,\alpha}(T, V, \{m_{i,\alpha}\}) \qquad \text{...... (6.9)}$$

A decrease of Helmholtz energy in a soil for which *T* is controlled by a thermal reservoir is equal to the reversible work delivered by the soil.

Gibbs Energy: With T, P and $\{m_{i,\alpha}\}$ as independent variables of state, the corresponding thermodynamic potentials are called Gibbs Energy (G).

$$dG = -SdT + VdP + \Sigma_i \Sigma_\alpha \mu_{i,\alpha} dm_{i,\alpha} \qquad \text{...... (6.10)}$$

and the corresponding state equations are:

$$S = S(T, P, \{m_{i,\alpha}\}), \quad V = V(T, P, \{m_{i,\alpha}\}),$$

$$\mu_{i,\alpha} = \mu_{i,\alpha}(T, P, \{m_{i,\alpha}\}) \qquad \text{...... (6.11)}$$

A decrease of Gibbs energy in a soil for which *T* and *P* are controlled by thermal and volume reservoirs is equal to the reversible work delivered by the soil.

The principal thermodynamic potentials are given in the Table 6.2.

Table 6.2: The principal thermodynamic potentials

Thermodynamic Potentials	Independent Variables of State	Symbols
Internal Energy	S, *V*, and $\{m_{i,\alpha}\}$	U(S, *V*, and $\{m_{i,\alpha}\}$)
Enthalpy	S, *P*, and $\{m_{i,\alpha}\}$	H(S, *P*, and $\{m_{i,\alpha}\}$)

Thermodynamic Potentials	Independent Variables of State	Symbols
Helmholtz Energy	T, *V*, and $\{m_{i,\alpha}\}$	A(T, *V*, and $\{m_{i,\alpha}\}$)
Gibbs Energy	T, *P*, and $\{m_{i,\alpha}\}$	G(T, *P*, and $\{m_{i,\alpha}\}$)

6.5 Thermodynamic Relationships

There are three important thermodynamic relationships derived from the fact that the total differentials of internal energy U and the thermodynamic potentials are exact and that U, H, A and G are extensive quantities. The relationships are:

(a) The Maxwell Relations.

(b) The Euler Equations and

(c) The Gibb's-Duhem Equation

6.5.1 The Maxwell Relations

This considers Gibb's energy equation as the exact differential. Then:

$$\frac{\partial^2 G}{\partial P \partial T} = \frac{\partial^2 G}{\partial T \partial P} \quad \text{... (6.12)}$$

And so on for all cross partial derivatives. However,

$$\frac{\partial^2 G}{\partial P \partial T} = \left(\frac{\partial S}{\partial P}\right)_{T,(m_{i\alpha})} \quad \text{... (6.13)}$$

$$\frac{\partial^2 G}{\partial T \partial P} = \left(\frac{\partial V}{\partial T}\right)_{P,(m_{i\alpha})} \quad \text{... (6.14)}$$

This results in

$$-\left(\frac{\partial S}{\partial P}\right)_{T,(m_{i\alpha})} = \left(\frac{\partial V}{\partial T}\right)_{P,(m_{i\alpha})} \quad \text{... (6.15)}$$

This relationship is example of Maxwell Relation which represents an equality between two cross partial derivatives of thermodynamic potentials. Similarly one can writes:

$$-\left(\frac{\partial S}{\partial_{m_{i\alpha}}}\right)_{T,P,\{m_{j\beta}\}} = \left(\frac{\partial \mu_{i\alpha}}{\partial T}\right)_{P,\{m_{i\alpha}\}} \quad \text{... (6.16)}$$

and

$$\left(\frac{\partial V}{\partial_{m_{i\alpha}}}\right)_{T,P,\{m_{j\beta}\}} = \left(\frac{\partial \mu_{i\alpha}}{\partial P}\right)_{T,\{m_{i\alpha}\}} \quad \text{...... (6.17)}$$

The principal utility of Maxwell relations is that they permits experimental alternatives in methods for measuring changes in thermodynamic quantities.

6.5.2 The Euler Equation

An Euler equation can be derived for the internal energy and for each thermodynamic potential based only on the fact that these quantities are extensive. Since Gibb's energy G is extensive, its mathematical properties are expressed as:

$$G_\alpha(T,P,\{\lambda m_{i\alpha}\} = \lambda G_\alpha(T,P,\{m_{i\alpha}\}) \quad \text{...... (6.18)}$$

Where, G_α is the Gibb's energy of the αth phase in a soil and $\lambda>0$ is a scale factor of arbitrary magnitude. By the chain rule of partial differentiation,

$$\frac{\partial G_\alpha}{\partial \lambda} = \sum_i \frac{\partial G_\alpha}{\partial_{xi\alpha}} \frac{\partial_{xi\alpha}}{\partial \lambda} = \sum_i \frac{\partial G_\alpha}{\partial_{xi\alpha}} m_{i\alpha} \quad \text{...... (6.19)}$$

Where, $x_{i\alpha} \equiv \lambda m_{i\alpha}$

Also, $$\frac{\partial G_\alpha}{\partial \lambda} = G_\alpha T,P,\{m_{i\alpha}\} \quad \text{...... (6.20)}$$

Or $$G_\alpha(T,P,\{m_{i\alpha}\}) = \sum_i \frac{\partial G_\alpha}{\partial x_{i\alpha}} m_{i\alpha} \quad \text{...... (6.21)}$$

After multiplying with

$$\lambda G_\alpha(T,P,\{m_{i\alpha}\} = G_\alpha(\mathrm{TP},\{x_{i\alpha}\}) = \sum_i \frac{\partial G_\alpha}{\partial x_{i\alpha}} x_{i\alpha} \quad \text{...... (6.22)}$$

Finally, λ is set equal to 1, gives the Euler equation

$$G_\alpha(T,P,\{m_{i\alpha}\} = \sum_i \frac{\partial G_\alpha}{\partial m_{i\alpha}} m_{i\alpha} = \sum_i \mu_{i\alpha} m_{i\alpha} \quad \text{...... (6.23)}$$

The Euler equation shows that the internal energy U and all of the thermodynamic potential can be expressed as sums of extensive variable being just its conjugate intensive variable. One variable in each term of the Euler equation is an independent variable of state and the other variable can be expressed as a function of all the independent variables of state.

6.5.3 The Gibbs-Duhem Equation

Let us consider Gibb's energy of a single phase α in a soil. The total differential of this thermodynamic potential is:

$$dG_{\alpha} = -S_{\alpha}dT + V_{\alpha}dP + \sum \mu_{i\alpha}dm_{i\alpha} \qquad \text{...... (6.24)}$$

The total differential of Gibbs energy of phase α is given in Euler equation is

$$dG_{\alpha} = \sum_{i}(\mu_{i\alpha}dm_{i\alpha} + m_{i\alpha}d\mu_{i\alpha}) \qquad \text{...... (6.25)}$$

These two equations are compatible only if the following relation holds identically:

$$S_{\alpha}dT - V_{\alpha}dP = \sum m_{i\alpha}d\mu_{i\alpha} = 0 \qquad \text{...... (6.26)}$$

This equation is called Gibbs-Duhem equation.

Gibbs-Duhem equation is an identity relationship among the differentials of the intensive variables that describe soil. This equation makes a general statement that, for a phase containing C components, the C and two intensive properties ofthe phase cannot be varied independently and only C and one of them can be varied.

Chapter - 7

Evapo-transpiration & Crop-Coefficient

7.1 Evapotranspiration

Evaporation is a thermodynamic process in which liquid water escapes from the surface of any object to the air in gaseous state at a temperature below the boiling point. It is also a process where heat energy present in the liquid water as latent heat goes away and the water body cools down which is called as evaporative cooling. This occurs from any surface which directly or indirectly takes part in hydrologic cycle. Evaporation that occurs from the plant surfaces is called transpiration. In transpiration water escapes from the plant body through the stomata. In presence of energy and a gradient of vapour pressure, this process continuously occurs from both soil surface and plant surface, hence the term evapo-transpiration is used to indicate the combined process in vegetated area. Evapo-transpiration (ET) is a representation of the evaporative demand of atmosphere, independent of crop growth and management factors (Allen *et al.* 1998). Evapotranspiration is the simultaneous process of transfer of water to the atmosphere by transpiration and evaporation in a soil system (Allen *et al.* 1998; Mavi and Tupper 2004). It is an important parameter for climatological and hydrological studies, as well as for irrigation planning and management (Sentelhas *et al.* 2010) as a major component of the agriculture water budget. Since it is difficult to separate evaporation and transpiration during crop growth, they are often expressed in one term ET. ET is expressed in linear unit.

Three conditions must be fulfilled for occurrence of evapo-transpiration i.e. availability of water from which evaporation or transpiration occur, presence of a vapour pressure gradient between the source and the sink and presence of energy for causing the transformation of liquid water into vapour. The important soil factor in evapotranspiration process is the presence of water at the root zone so that water can move from the soil to the leaves to meet the transpiration demand which in turn depends on certain plant related factors. The important plant related factors are:

a. Rooting characteristics of the plant, since this is where water is taken up by the roots from the soil through the process of osmosis.

b. The physiological process and other modification to overcome the impediments for flowing water from the root zone to the stomata under both atmospheric and soil moisture stress.

c. Canopy characteristics i.e. relative distribution of stomata, stomatal density at both upper and lower surfaces of the leaves.

d. Aerodynamic characteristics of the leaves in relation to the atmosphere around it and lastly

e. Three dimensional nature of the evaporating medium.

The transfer of water from the soil and plant to atmosphere is a relationship of source and sink and basically controlled by environmental factors such as atmospheric temperature, net radiation, soil heat flux and the wind velocity that set in the energy gradient between the source (soil and plant) and the atmosphere. It is a function of weather parameters, though independent, largely influences on crop productions.

The knowledge of evapotranspiration is paramount for water management in crops. Water is lost through transpiration process and only 1% of total water taken up by the crop through its root system is used for crops' metabolic activities. The amount of water taken up by the plants should be supplied to the soil system through irrigation in order to provide a stress free conditions to the plant for their growth. With the knowledge of evapotranspiration, therefore, one can decide the amount of water to be supplied to the crops through irrigation. Since evapotranspiration is a continuous process, one can also determine the rate of loss of water from the crop fields with the information on evapotranspiration.

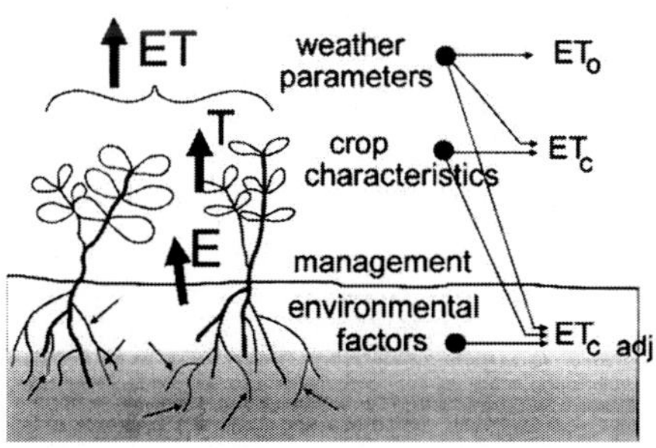

Fig. 7.1: Depiction of Evapotranspiration Process

7.2 Determination of Evapotranspiration

The combined loss of water from soil through evaporation and from crop canopy through transpiration is difficult to determine in the field. Neither can they be easily separated out. Further, the loss of water from water surface can be easily determined with the help of Open Pan Evaporimeter installed in any weather station. However, evaporation from water surface cannot be similar to the evapotranspiration from cropped field. The energy depth in the soil profile and the transpiration process are quite different from the process of evaporation. Many workers use the data of Pan Evaporimeter to determine the evaporation from the cropped field by considering Pan Coefficients (K_p) and Crop Coefficients (K_c). But this method is found to be highly approximated one. Determination of crop evapotranspiration can be done only with an instrument called Lysimeter where daily loss of water from the root zone can be determined with the help mass balance in the instrument.

7.2.1 Lysimeter

The name Lysimeter is of Greek origin. 'Lysis' means dissolution or movement and 'metron' means measurement. This means Lysimeter is an instrument used to measure the movement of fluid. It is in fact a multifaceted instrument used for variety of investigations particularly in the study of water balance and of percolation of water beneath the root zone. While studying the water balance, it is obvious that the different component of basic water balance equation can be studied in this instrument. The evaporative fraction of water balance is precisely measured through precision Lysimeter and hence widely used in the study of crop evapotranspiration.

Lysimeters are basically of two types- weighing type lysimeter and non-weighing type lysimeter. Non-weighing types are generally drainage type lysimeters where only drainage water collection is the main objective. Weighing type lysimeter weighs the change in the storage water volume thereby determines the different losses through partitioning. Typical gravitation or weighing type lysimeters are extensively used in Europe. Recent development of various sensor technologies particularly, moisture sensors, suction collectors, etc. along with the data logger, Lysimeter technique has got renewed importance in research arena. Generally, gravitational lysimeter in Europe has one m^2 surface area and total depth of 2 m. About 15 cm thick filter layer is placed below the lysimeter to avoid upward fluxes of groundwater. Presently, lysimeters are equipped with precision load cell, TDR probes, Tensiometer and vacuum controlled suction apparatus to collect gravitational water under pre-determined suction.

In-situ monitoring of water, its dynamics and solute flux is very challenging task. The basic characteristics regarding the controlling factors are also very complex in nature. Further, the question of sustainability, environmental degradation, leachate in relation to the soil types and genesis are not always quantitative. Exact quantification of soil water flow is however necessary to predict solute transfer within the unsaturated zone where saturated hydraulic conductivity does not have any role. Different methods are exiting for indirect measurement of the solute transfer including the mathematical and analytical models developed for both saturated and unsaturated zones.

Both direct and indirect methods are recommended by Gee *et al.* (2002). The simplest form of indirect estimation of drainage is the use of basic water balance equation, which can be written as:

$$P + I = ET + \Delta S + D \pm RO \qquad \text{...... (7.1)}$$

Where, P= Precipitation

I = Irrigation

ET = Evapotranspiration

ΔS = Change of storage

D = Drainage water

RO= Run off or Run on

In absence of irrigation and runoff and for periods when storage change is negligible drainage is typically determined as the difference between precipitation and ET. Here ET is estimated from the meteorological parameters.

Meteorological parameters are generally available for most of the location with the network of the Automatic Weather Stations (AWS) under IMD, Govt. of India. Therefore, water balance-ET estimation method has been widely used. In humid region the evaporative demand is less than the precipitation and can be obtained with better accuracy. In arid region, the accuracy is many time less.

Another widespread indirect method is measurement of specific soil characteristics and calculating water flux rate using data from Tensiometer, Time domain Reflectometry (TDR) or Heat Pulse Probe (Gee *et al.* 2002).

The most important direct measurement is by any buried equipment which can collect drainage water - can be termed as drainage lysimeter. A wide range of lysimeter with simple water collector have been used ranging from small free-draining pan type lysimeter, wick sampler or tension controlled lysimeter (Meibner *et al.* 2000; Gee *et al.* 2009; Aboukhaled *et al.*

1982; Meibner *et al.* 2008). Only lysimeters are capable of direct measurement of drainage water or the water percolating through soil profile. If the lysimeter is weighable they give the evapo-transpiration also. A large lysimeter is suitable for obtaining accurate data on quantity and quality of drainage water.

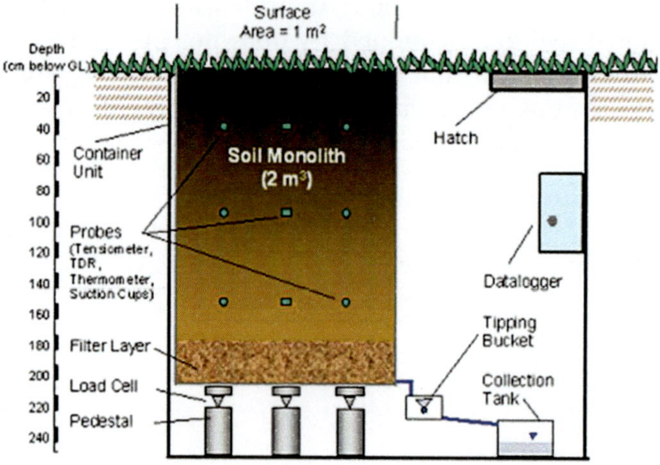

Fig. 7.2: Schematic diagram of a weighing type lysimeter (Meissner and Seyfarth, 2004)

7.2.2 Installation of Lysimeter

Installation of large lysimeter is a great challenge. Installing the lysimeter by excavating the soil may perhaps be the easiest option but it disturbs the soil physical properties. Disturbed soil samples in lysimeter do not give the exact information because of the change of soil porosity, compactness and profile orientation. Further, disturbed soil sample, even if the profile orientation is maintained, takes long time to settle down to have the original soil compactness. Hence, insertion of monolithic soil sample into the lysimeter or insertion of lysimeter into the soil and separating it from the parent profile later on is the new technique in the lysimetry. The newly developed cutting tools made it possible to cut monolithic soil sample for the lysimeter. The soil monolith is not damaged during the cutting process. The extraction of soil in minimally affected since due to reducing coat friction the force required to cut the monolith is small. This technique can be used for different soil type (Meibner *et al.* 2007). After insertion of the soil monolith, the probes *viz.* TDR based moisture probe, tensiometers, vacuum based drainage water collector, temperature probe etc. are installed at different depths to represent the suitable root zone of the particular crop. All the probes then connected to data logger for continuous collection of the different information of relevant parameters. An AWS should also be installed in the lysimeter site for collection of all the weather parameters to estimate the reference ET based on suitable model, which is generally FAO recommended Penman-Montieth model (FAO Irrigation and Drainage Paper No. 56, 1992).

Fig. 7.3: Cutting tool for lysimeter (Meissner and Seyfarth, 2004)

Fig. 7.4: Installation of different probes at suitable depth (UMS-GmBH Lysimeter Manual)

7.2.3 Weighing Precision

In water balance studies, lysimeters are measuring the quantity of rainfall, actual evapo-transpiration and drainage. High precision lysimeter can also be capable of measuring the precipitation from dew and fog. Present lysimeters are generally high precision due to invent of different kind of probes, high temporal resolution is also possible which can be helpful to modelling hydrologic process.

In case of water balance studies in crops, Lysimeter should replicate the actual field conditions where the crop is grown. The evaporation from the soil and the transpiration from the crop canopy should represent exactly that is occurring in the field. Therefore, the cultural management practices including water management should be similar in the both lysimeter and the filed. Management but not the physical process is at the control of researchers. Therefore, a reference Tensiometer is also installed in the field which will control the suction in the lysimeter i.e. the vacuum pump will take the reference suction from the field Tensiometer to create equivalent suction to the vacuum pump so that the soil moisture status at any given time will remain same in the field and the lysimeter.

7.2.4 Drum Culture Technique

Another method is used by the field workers in absence of Lysimeter, since Lysimeter is found to be a costly equipment if one goes for high precision results. Drum culture technique can not only determine the evaporation but also can separate the evaporation and transpiration. In this method four drums are taken for the study. Two drums out of the four should be provided with perforations to facilitate drainage where the other two will have blind bottoms. All the drums should be filled with soils from the field where the experiment is conducted and maintained the same soil profile as in the field. All the drums should have the provision to weight with the help of balance, so that difference of water can be measured on daily basis.

Two drums should be planted with the crops and the other two should be left without crops. The drums are irrigated with equal amount irrespective the presence of crops or not. In the blind bottom drum (Fig. 8.5a), as it has only soil without perforation at the bottom, the loss of water can be possible only through evaporation. In the second drum (Fig. 5.5b), which is also blind at the bottom, as crop is grown there, both evaporation and transpiration take place. The water lost from this drum is only through evapotranspiration. The result from the first and second drum may not be accurate as there may be some gravitational water in the soil profile which would have drained out during the course of observation. The same study now repeated in two more drums with perforated bottoms.

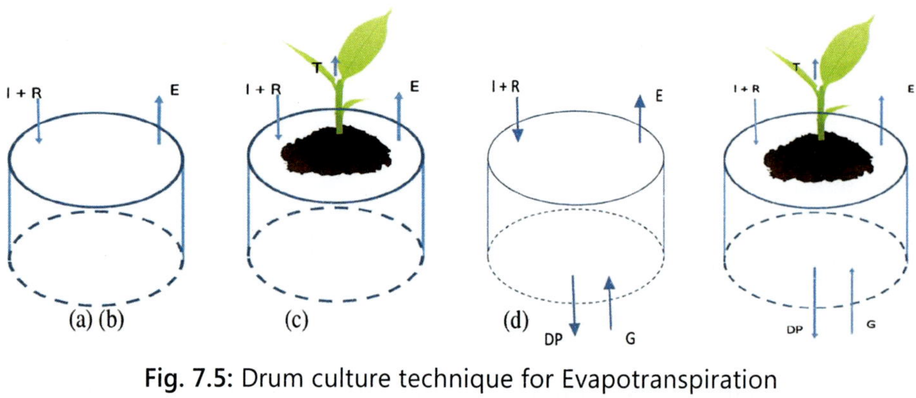

Fig. 7.5: Drum culture technique for Evapotranspiration

Though ground water contribution through capillary rise can be an issue, this may be avoided by installing the drum at the place where water table presence is below two meters from the root zone. Considering the, negligible water rise from the ground water table to the drum, the loss of water in both the drums take place though deep percolation apart from E and T. By taking observation in 3rd and 4th drums, all the water component going out of the soil profile i.e. E, T and DP can be determined. This method is though devoid of sophistication, but gives fairly good result.

Direct measurement of evapotranspiration is crop specific and event specific which cannot be generalised across all the crops and places. Use of lysimeter is also a costly affair. Hence, models are being used for estimation of evapotranspiration.

7.3 Evapo-transpiration Models

It has already been discussed that evapotranspiration is influenced by weather parameters and canopy characteristics apart from water availability in the root zones in the soil. Weather parameters, particularly the bright Sun shine hours are dependent on the day of the year and the latitude. There are six categories of models available for evapotranspiration estimation. These are:

a. Water budget method
b. Aerodynamic method
c. Energy budget method
d. Eddy correlation technique
e. Boundary layer and stomatal approach and
f. Combination method

Some books has the classification of the models as follows:

a. Temperature dependant methods
b. Radiation dependant methods and
c. Combination methods.

We shall look into few very commonly used methods for estimation of evapo-transpiration.

One should underscore the difference between the Potential Evapo-transpiration and Reference Evap-transpiration. Potential Evapo-transpiration is estimated from the crop surfaces under non-restrictive conditions having optimum water supply. The models like Blaney-Criddle, Thornthwaite, Turc, and Jensen and Haise estimate potential evapo-transpiration. Potential Evapo-transpiration models though give estimates of evapo-transpiration at times fairly acceptable level, one should understand that these models do not includes the plant leaves surfaces or

LAI. The crop specific or leaf surface specific evapo-transpiration in fact is not estimated. These models depend on three basic premises:

a. The water is continuously and adequately supplied.

b. Energy to change liquid water to water vapour and

c. A vapour gradient to maintain the flux from the evaporating surfaces to the atmosphere.

Various models are developed based on one or more of the aforesaid requirement. The soil-water balance models are based on the first condition and the energy balance models are based on the second condition. However, the combinations methods are dependent on the conditions two and three. Some of the models available in the literatures are listed below:

Mass Transfer Based Models:

The mass transfer based models used the concept of eddy transfer of water vapour from the evaporating surface to the atmosphere. The fundamental principle of Dalton's Gas Law of partial pressure is employed here. This is essentially dependent on the partial vapour pressure based transfer of water vapour from the surface of the leaves to the atmosphere. Many models give satisfactory estimation of evapo-transpiration. Some of the mass transfer based models are listed below in Table 7.1 to 7.4:

Table 7.1: Mass Transfer models

Dalton	$ET_o = (0.3648 + 0.07223\ u) \times (e_s - e_a)$
Trabert	$ET_o = 0.3075\ (u)^{1/2} \times (e_s - e_a)$
Meyer	$ET_o = (0.375 + 0.05026u) \times (e_s - e_a)$
Albrecht	$ET_o = (0.1005 + 0.2974u) \times (e_s - e_a)$
Romanenko	$ET_o = 0.0018 \times (T_a + 25)^2 \times (100\text{- RH})$
Brockamap and Wenner	$ET_o = 0.543 \times u^{0.456} \times (e_s - e_a)$
WMO	$ET_o = (0.1298 + 0.0934u) \times (e_s - e_a)$
Mahringer	$ET_o = 0.15072 \times (3.6u)^{1/2} \times (e_s - e_a)$

Where u= Wind velocity, m/s and e_s =Saturation vapour pressure, hPa, e_a = Actual vapour pressure, hPa

Table 7.2: Temperature Based Models

Blaney-Criddle	$ET_0 = c[p(0.46T+8)]$ mm/ day Where p= percentage of total annual day time hour obtained from table.
Thornthwaite Formula (PET)	$E_p = 1.6\left(\frac{10t}{TE}\right)^a \frac{N}{12}$ a= $0.000000675(TE)^3 - 0.0000771\,(TE)^2 + 0.01792TE + 0.49239$ $TE = \sum_{j=1}^{12}\left(\frac{t_j}{5}\right)^{1.514}$ $N = 2\cos^{-1}(-\tan\phi\tan\delta)\frac{12}{\pi}$
Schendel	$ET_0 = 16\times\frac{T_a}{RH}$
Hargreaves 1	$ET_0 = 0.408 \times 0.0030 \times (T_a + 20) \times (T_{max} - T_{min})^{0.4} \times R_a$
Hargreaves 2	$ET_0 = 0.408 \times 0.0025 \times (T_a + 16.8) \times (T_{max} - T_{min})^{0.5} \times R_a$
Hargreaves 3	$ET_0 = 0.408 \times 0.0013 \times (T_a + 17) \times (T_{max} - T_{min} - 0.123P)^{0.76} \times R_a$
Hargreaves 4	$ET_0 = 0.408 \times 0.0023 \times (T_a + 17.8) \times (T_{max} - T_{min})^{0.424} \times R_a$

Table 7.3: Radiation Based Models

IRMAK 1	$ET_0 = -0.611 + 0.149 \times R_s + 0.079\,T_a$
IRMAK 2	$ET_0 = -0.642 + 0.174 \times R_s + 0.0353\,T_a$
IRMAK 3	$ET_0 = -0.478 + 0.156 \times R_s - 0.0112\,T_{max} + 0.0733\,T_{min}$
JONES & RITCHIE	$ET_0 = \alpha\,[3.87 \times 10^{-3} \times R_s \times (0.6\,T_{max} + 0.4 \times T_{min} + 29)]$ $5\,°C < T_{max} < 35\,°C$, $\alpha = 1.1$ $T_{max} > 35\,°C$, $\alpha = 1.1 + 0.05 \times (T_{max} - 35)$ $T_{max} < 5\,°C$, $\alpha = 0.1 \times e^{[0.18 \times (Tmax + 20)]}$
JENSEN HAISE	$ET_0 = C_T\,(T_a - T_x) \times R_s$; where $C_T = \frac{1}{\left[\left(45 - \frac{h}{137}\right) + \left(\frac{365}{e^{T_{max}} - e^{T_{min}}}\right)\right]}$ $T_x = -2.5 - 0.14 \times (e^{T_{max}} - e^{T_{min}}) - \frac{h}{500}$

RAVAZZANI	$ET_0 = (0.817 - 0.00022 \times Z) \times 0.0023 \times R_a (T_a + 17.8) \times (T_{max} - T_{min})^{0.5}$

Energy Balance Approach:

The energy balance technique consists of dividing the energy transferred into atmosphere into sensible and latent heat components by measuring the corresponding vertical differences in temperature and humidity above the surface.

Table 7.4: Combination models

Penman	$E_0 = 0.35 \{1 + 0.98/ (100u)\} x (e_s - e_a)$ Where, u (wind velocity) is in mm/d and e (vapour pressure) is in mm of Hg
Rohwer	$ET_O = 0.44 \times (1 + 0.27u) x (e_s - e_a)$ Where, u (wind velocity) is in miles/d and e (vapour pressure) is in mm of Hg

7.4 Development of Reference Evapo-transpiration Model

Penman (1948) measured evaporation from open water surface and then used that reference evaporation. With the multiplication of crop-coefficient he determined crop evapo-transpiration. The approach is still used by many field workers for designing irrigation systems in absence of accurate evapo-transpiration data which is location and crop specific. Doorenbos-Pruitt (1977) proposed a modified form of Penman method of evapo-transpiration and used reference crop surfaces instead of open water surface. They defined reference surface as:

> *"An extended surface of an 0.08 to 0.15 m tall green grass cover of uniform height, actively growing, completely shading the ground and not short of water".*

Many researchers however reported that Modified Penman method over predicted evapo-transpiration. Using the same principle, Monteith later developed an equation that described the transpiration from a dry, extensive, horizontal and uniformly vegetated surface, fully covering the ground that is optimally supplied with water. This equation is known as Penman-Monteith equation.

In Penman-Monteith equation the grass reference crop is defined as:

> *"A hypothetical crop fully covering the ground, and not short of water with an assumed crop height of 0.12 m, a fixed canopy resistance of 70 s/m and a canopy reflection coefficient of 0.23."*

This reference surface is widely accepted by the researchers and the field workers for all practical purposes.

Allen *et al.* (2005a) reported that the FAO Penman-Monteith (FAO PM) method had been considered as a universal standard to estimate ET called as reference ET (ET_0) and it incorporates physiological and aerodynamic parameters at location specific observation (Allen *et al.*1989, Allen and Pruitt 1991, Lopez-Urrea *et al.* 2006). It provides consistent ET_0 in many regions and climates. This method has been accepted worldwide as a good estimator compares to others method especially for daily computations (e.g. Pereira and Pruitt 2004, Cai *et al.* 2007, Stöckle *et al.* 2004, Harmsen 2003, Mohan and Arumugam 1996, Itensifu *et al.*, 2003, Li *et al.*, 2003, Tyagi *et al.* 2000; Chiew *et al.* 1995). In other words, transpiration slowly supplants evaporation as Crop Evapo-transpiration (ET_c) increases. Soil water availability encourages plant growth and subsequent increases in ET_O. However, deficient soil water can induce plant wilting or death (Brady, 1990).

Study on ET_0 was conducted by Tyagi *et al.* (2000) on rice (monsoon) using regression statistics between Penman-Monteith and different methods where there was better agreement between Penman Monteith and FAO Blaney-Cridle (FB-C) along with FAO-24 (FAO Irrigation and Drainage Paper No. 24) corrected Penman followed by other methods and the higher value was estimate by FAO -24 corrected Penman. Another study was done on high lands of Eastern Ghats of Orissa, where comparison of some empirical methods for estimation of ET_0 was conducted to evaluate the best method for estimation of ET_0. Results indicated that FAO-56 (FAO Irrigation and Drainage Paper No. 56) was best in estimation of ET_0 throughout the year and Hargreaves equation being more consistent among temperature based equations (Lenka *et al.* 2009). Pictorial representation of estimation of ET_0 and ET_c are shown in Table 7.6.

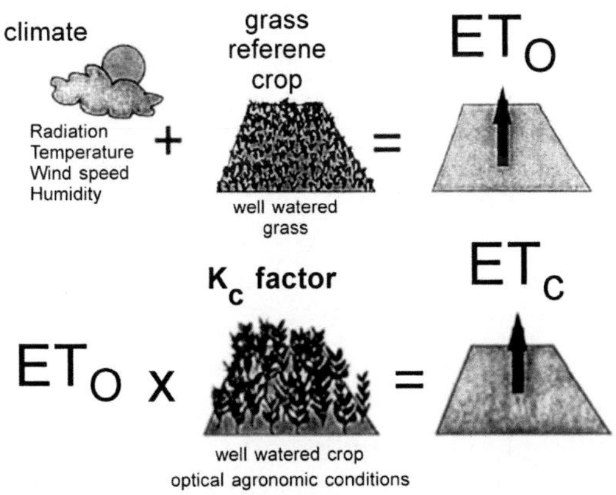

Fig. 7.6: Pictorial representation for estimation of ET_O and ET_c

The actual crop water use depends on climatic factors, crop type and crop growth stage. While ET_0 provides the climatic influence on crop water use, the effect of crop type and management is addressed by ET_c. Factors affecting ET_c such as ground cover, canopy properties and aerodynamic resistance for a crop are different from the factors affecting reference crop (grass or alfalfa); therefore, ET_c differs from ET_O. The characteristics that distinguish field crops from the reference crop are integrated into as crop factor or crop coefficient (k_c) (Allen *et al.* 1998). k_c is used to determine the actual water use for any crop in conjunction with ET_O.

$$ET_c = k_c \times ET_0 \qquad \text{...... (7.2)}$$

In general it is difficult to determine the ET_c as a residual from water balance computation, so k_c based approach is primary for predicting water consumption from irrigation projects (Burt *et al.* 1997; Molden and Sakthivadivel 1999; Droogers and Bastiaanssen 2002).

Crop coefficients (k_c) are properties of plants used in predicting evapo-transpiration (ET). It varies by crop, stage of growth of the crop, and by some cultural practices. Crop coefficients are crop and crop variety specific and also location specific. Doorenbos and Pruitt (1977) recommended the values of crop coefficients at different stages of growth i.e. initial, crop development, mid season and maturity under different RH and Wind Speed conditions. The k_c values recommended by FAO -.24 (Doorenbos and Pruitt 1977) were used worldwide to estimate actual evapo-transpiration based on the local weather parameters in absence of derived crop coefficients. But if crop duration and morphology do not match, the actual values differ considerably from the tabulated values. Additions of more crops were done in the list of FAO-56 by Allen *et al.* (1998). They recommended the values for broad climatic conditions indicating maximum height of the crops for use of specific sets of crop coefficients. Crop-coefficients are generally reported in literature for three distinct growth stages *viz.* initial (development stage), mid (maximum vegetative growth) and end (senescence stage).

The initial period is defined in FAO-24 (Doorenbos and Pruitt 1977) and in FAO-56(Allen *et al.* 1998) for annual crops as the period between the planting date and the date of approximately 10% ground cover. This period represents conditions when the soil is effectively bare. If the soil surface is wet during this period, the evaporation rate may be relatively large. As the soil surface dries, hydraulic conditions change and evaporation decreases. The mean crop coefficients (k_c) during this period are termed the crop coefficient for the initial period k_{cini}. A value for k_{cini} is required for constructing a "singular" crop coefficient curve for a growing season that incorporates impacts of wetting frequency on k_c. FAO-56 included a "dual"

k_c method in addition to the singular method that simulates impacts of evaporation separately (Allen *et al.,* 2005). However, the singular k_c method is often applied for general planning studies and regional analyses and thus, accurate and representative k_{cini} values are needed.

7.5 Penman-Monteith Method and its Calculation procedure

ASCE-EWRI (2005) adopted the same definition of reference surface for standardized ET_0 with a provision for lower surface day time resistance of 50 s/m and higher night time resistance of 200 s/m, when calculating on hourly or shorter time steps. ASCE-EWRI resistance for hourly calculations were later adopted by FAO.

The ASCE-EWRI (2005) standardized Penman-Monteith equation is given as:

$$ET_O = \frac{0.408\Delta(R_n - G) + \gamma \dfrac{C_n}{T+273} u_2 (e_s - e_a)}{\Delta + \gamma(1 + C_d u_2)} \quad \text{...... (7.3)}$$

Where,

ET_0 = Standardized reference evapo-transpiration, mm/h or mm/d

R_n = Calculated net radiation at the crop surface, $MJ/m^2/day$ or $MJ/m^2/h$

G = Soil heat flux density, $MJ/m^2/day$ or $MJ/m^2/h$

T = mean daily or hourly air temperature at 1.5 to 2.5 m height, °C

u_2 = Daily or hourly wind velocity at 2 m height, m/s

e_s = saturation vapour pressure at 1.5 - 2.5 m height (kPa) calculated for daily time steps

e_a = mean actual vapour pressure at 1.5-2.5 m height (kPa)

Δ = slope of the saturation vapour pressure and temperature curve, kPa/°C

γ = Psychrometric constant, kPa/^{0}C

C_n = Numerator constant that changes with reference type and calculation time step (K mm s^3/Mg/h)

C_d = Denominator constant changes with reference type and calculation time step (s/m)

The constant term 0.408 bears the unit of (m^2 mm/MJ)

The values of C_d and C_n are given in the Table 7.5 for standardized ET_O.

Table 7.5: Values of C_d and C_n for Eq. No. 5.3 (after Allen *et al.* 2005)

Calculation time step	Short Reference ET_O (clipped grass)		Unit for ET_O	Unit for R_n, G	G/R_n Ratio
	C_n	C_d			
Daily	900	0.34	mm/day	MJ/m²/day	0
Hourly during day time	37	0.24	mm/h	MJ/m²/h	0.1
Hourly during night time	37	0.96	mm/h	MJ/m²/h	0.5

7.6 Step by Step Computation of Parameters in P-M Equation (Zotarelli *et al.* 2010)

Step 1:

To determine mean daily temperature, the (average) daily maximum and minimum air temperatures in degrees Celsius (°C) are required. Where only (average) mean daily temperatures are available, the calculations can still be executed but some underestimation of ETo will probably occur due to the non-linearity of the saturation vapor pressure - temperature relationship (Allen *et al.* 1998).

Average temperature is calculated by:

$$T_{mean} = \frac{T_{\min} + T_{\max}}{2} \quad \text{...... (7.4)}$$

Where, T_{mean} = mean daily air temperature, °C; T_{max} = maximum daily air temperature, °C; T_{min} = minimum daily air temperature, °C.

Step 2: Mean daily solar radiation (R_s)

The average daily net radiation expressed in megajoules per square meter per day (MJ m^{-2} day^{-1}) is required. A simple average of solar radiation values obtained from a weather station in the period of 24h (0:00:01 am to 11:59:59 pm) is required. The conversion of units may be required when solar radiation is expressed in watts per square meter per day (W m^{-2} day^{-1}).

$$R_s \ (\text{MJ m}^{-2}\text{day}^{-1}) = \text{Rs} \ (\text{W m}^{-2}\text{day}^{-1}) * 0.0864 \quad \text{...... (7.5)}$$

Step 3: Wind speed (u_2)

The average daily wind speed in meters per second (m s^{-1}) measured at 2 m above the ground level is required. It is important to verify the height at which wind speed is measured, as wind speeds measured at different heights above the soil surface differ. The wind speed measured at heights other than 2 m can be adjusted according to the follow equation:

$$u_2 = u_h \frac{4.87}{\ln(67.8 - 5.42)} \qquad \text{...... (7.6)}$$

Where,

u_2 = wind speed 2 m above the ground surface, m s^{-1} ;

u_z = measured wind speed z m above the ground surface, m s^{-1} ;

h = height of the measurement above the ground surface, m.

In case of wind speed is given in miles per hour (mi h^{-1}) the conversion to m s^{-1} is required. u_2 (m s^{-1}) = u_2 (mi h^{-1}) * 0.447

Step 4: Slope of saturation vapor pressure curve (Δ)

For the calculation of evapotranspiration, the slope of the relationship between saturation vapor pressure and temperature, Δ, is required.

$$\Delta = \frac{4098\left[0.6108\ \exp\left(\frac{17.27\ T_{mean}}{T_{mean} + 237.3}\right)\right]}{(T_{mean} + 237.3)^2} \qquad \text{...... (7.7)}$$

Where, T_{mean} = mean daily air temperature, °C,

exp = 2.7183 (base of natural logarithm).

Step 5: Atmospheric Pressure (P)

The atmospheric pressure, P, is the pressure exerted by the weight of the earth's atmosphere. Evaporation at high altitudes is promoted due to low atmospheric pressure. This effect is, however, small and in the calculation procedures, the average value for a location is sufficient. A simplification of the ideal gas law, assuming 20°C for a standard atmosphere, can be employed to calculate P in kPa at a particular elevation:

$$P = 101.3\left[\frac{293 - 0.0065z}{293}\right]^{5.26} \qquad \text{...... (7.8)}$$

Where, z = elevation above mean sea level, m

Step 6: Psychrometric constant (γ)

The psychrometric constant (γ) relates the partial pressure of water in air to the air temperature so that vapor pressure can be estimated using paired dry and wet thermometer bulb temperature readings. Another way to describe the psychrometric constant is the ratio of specific heat of moist air at constant pressure (C_p) to latent heat of vaporization. The specific heat at constant pressure is the amount of energy required to increase the temperature of a unit mass of air by one degree at constant pressure. Its value depends on the composition of the air, i.e., on its humidity. For average atmospheric conditions a C_p value of 1.013 10^{-3} MJ kg^{-1} °C^{-1} can be used. As an average atmospheric pressure is used for each location, the psychrometric constant is kept constant for each location depending of the altitude:

$$\gamma = \frac{CpP}{\varepsilon\lambda} = 0.000665P \qquad \text{...... (7.9)}$$

Where, γ = psychrometric constant, kPa °C^{-1}

P = atmospheric pressure, kPa

λ = latent heat of vaporaization, 2.45 $MJkg^{-1}$

C_p =specific heat at constant pressure, 1.013 10^{-3} MJ kg^{-1} °C^{-1}

ε = ratio molecular weight of water vapour/dry air= 0.622

Step 7: Delta Term, DT (Auxiliary calculation for Radiation term)

In order to calculate the radiation term, several other terms are to be calculated. These are particularly Delta Term (DT) and Psi Term (PT). The DT is calculated as follows:

$$DT = \frac{\Delta}{\Delta + \lambda(1 + 0.34\,u_2)} \qquad \text{...... (7.10)}$$

Terms are as used earlier.

Step 8: Psi term (Auxiliary calculation for wind term)

The latitude term is calculated as follows:

$$PT = \frac{\Delta}{\Delta + \lambda(1 + 0.34\,u_2)} \qquad \text{...... (7.11)}$$

Terms are as used earlier.

The wind term also needs temperature term (TT) which is calculated as:

$$TT = \left[\frac{900}{T_{mean} + 273}\right] \times u_2 \qquad \text{...... (7.12)}$$

Step 9: Mean saturation vapour pressure derived from air temperature (e_s)

The e_s can be calculated from the air temperature as it is related to the temperature of air. The relationship is given as :

$$e_s = 0.6108\ exp\left[\frac{17.27T}{T + 237.3}\right] \qquad \text{...... (7.13)}$$

Where, e_s = Saturation vapour pressure in kPa; T= air temperature in °C

Saturation vapour pressure is calculated here with mean air temperature. The same can be calculated by using T_{max} and T_{min} and then calculating the average of the value.

$$e_s(T_{max}) = 0.6108\ exp\left[\frac{17.27T_{max}}{T_{max} + 237.3}\right] \quad \text{...... (7.14)}$$

and

$$e_s(T_{min}) = 0.6108\ exp\left[\frac{17.27T_{min}}{T_{min} + 237.3}\right] \quad \text{...... (7.15)}$$

And the

$$e_s = \frac{e_{s(Tmax)} + e_{s(Tmin)})}{2} \quad \text{...... (7.16)}$$

Step 10: Actual vapour pressure derived from Relative Humidity (e_a)

Depending upon the availability of humidity data, the actual vapour pressure can be calculated either from mean RH or from maximum and minimum relative humidity. The relationship for calculation of actual vapour pressure is given below:

$$e_a = \frac{e_{s(Tmax)}\left(\frac{RH_{max}}{100}\right) + es(T_{min})\left(\frac{RH_{min}}{100}\right)}{2} \quad \text{...... (7.17)}$$

Terms are as used in earlier equation.

If the measurement of RH_{min} is considered erroneous or in doubt, one can use only RH_{max} for calculation of e_a:

$$e_a = e_{s(Tmin)}\left[\frac{RH_{max}}{100}\right] \quad \text{...... (7.18)}$$

In absence of RH_{min} and RH_{max}, one can use the RH_{mean} for calculation of e_a:

$$e_a = \frac{RH_{mean}}{100}\left[\frac{e_{s(Tmin)} + e_{s(Tmax)}}{2}\right] \quad \text{...... (7.19)}$$

Sometimes, RH data are not available or questionable. In that case e_a can be obtained by assuming when the air temperature is close to T_{min}, the air is nearly saturated with water vapour and the relative humidity is 100%. In other words, the dew point temperature is near the daily minimum temperature. If T_{min} is considered equal to T_{dew}, then e_a is equal to $e_{s(Tmin)}$.

Step 11: Calculation of inverse earth-sun relative distance (dr) and solar declination (δ) from Julian day.

In order to calculate solar radiation, the inverse relative distance between the earth and the sun is calculated along with the solar declination from Julian day.

$$d_r = 1 + 0.033\cos\left[\frac{2\pi}{365}J\right] \quad \text{...... (7.20)}$$

and

$$\delta = 0.409\sin\left[\frac{2\pi}{365}J - 1.39\right] \quad \text{...... (7.21)}$$

Where J is the number of day in the year between 1 to 365 or 366 as the case may be.

Step 12: Conversion of Latitude degrees to Radians

The solar radiation is dependent on the latitude of the location. However, in the equation the latitude is not taken as degrees but as radians. It is converted by the following relationship.

$$\varphi(radian) = \frac{\pi}{180}\varphi(decimal\ \text{degree}) \quad \text{...... (7.22)}$$

Step 13: Sunset hour angle (ω_s)

The sunset hour angle gives the day length of the area. This can be calculated as follows:

$$\omega_s = arc\cos[-\tan(\varphi)\tan(\delta)] \quad \text{...... (7.23)}$$

Where φ is in radians and δ as calculated.

Step 14: Extra-terrestrial radiation (R_a)

The extra-terrestrial radiation is an important parameter in the estimation of net radiation affecting the evapo-transpiration. The extra-terrestrial radiation in a location can be estimated from solar constant, solar declination and the time in a year.

$$R_a = \frac{24(60)}{\pi}G_{SC}d_r[(\omega_s\ \sin\varphi\sin\delta) + \cos\varphi\cos\delta\sin\omega_s] \quad \text{......(7.24)}$$

Where,

R_a = Extra-terrestrial radiation, MJ/m²/day

G_{SC} = Solar constant, 0.0820 MJ/m²/day

Other terms are as mentioned earlier.

Step 15: Clear sky solar radiation (R_{so})

Clear Sky radiation can be calculated with the following equation:

$$R_{SO} = (0.75 + 2E10^{-5} z)R_a \qquad \text{...... (7.25)}$$

Where, z = elevation from mean sea level

Step 16: Net solar or Net Shortwave radiation (R_{ns})

The net short wave radiation is the resultant radiation of net incoming and reflected solar radiation and estimated with the following equation.

$$R_{ns} = (1 - a)R_s \qquad \text{...... (7.26)}$$

Where, a = albedo, which is 0.23 for the reference grass surface

R_s = Incoming solar radiation, MJ/m²/day

Step 17: Net out going long wave solar radiation (R_{nl})

Stefan-Boltzmann law states that the rate of long wave energy emission is proportional to the absolute temperature of the surface raised to the fourth power. The net energy flux leaving the earth surface is however, less than the emitted and given by Stefan-Boltzmann law due to the absorption and downward radiation from the sky. Water vapour, clouds, CO_2 and dust are absorbers and emitters of long wave radiation.

It is thereby assumed that the concentrations of the other absorbers are constant:

$$R_{nl} = \sigma\left[\frac{(T_{max} + 273.16)^4 + (T_{min} + 273.16)^4}{2}\right]\left(0.34 - 0.14\sqrt{e_a}\right)\left[1.35\frac{R_s}{R_{so}} - 0.35\right] \qquad \text{...... (7.27)}$$

Where, R_{nl} = Out going long wave radiation, MJ/m²/day

σ = Stefan-Boltzmann constant (4.9003×10^{-9} MJ/K⁴/m²/day)

Temperature term is in Kelvin

Step 18: Net Radiation (R_n)

The net radiation is the balance of net incoming short wave radiation and net out going long wave radiation.

$$R_n = R_{ns} - R_{nl} \qquad \text{...... (7.28)}$$

The net radiation (R_n) is expressed in equivalent of evaporation (R_{ng}) in 'mm':

$$R_{ng} = 0.408 \times R_n$$

However, in the Penman-Monteith Equation, R_{ng} term is not used. In stead, $0.408 \times R_n$ is directly used in the equation.

In auxiliary calculation as stated earlier, ET_O is calculated directly with the help of radiation term and wind term.

Radiation Term $ET_{rad} = DT \times R_{ng}$ ...... (7.29)

Where, ET_{rad} is in mm/day.

Wind Term $ET_{wind} = PT \times TT\,(e_s - e_a)$ ...... (7.30)

Hence, $ET_O = ET_{rad} + ET_{wind}$ ...... (7.31)

Chapter - 8

Drought Indices for Drought Characterization

8.1 Introduction

The impact of climate change on the eco-system is looming large. It is affecting specially the water system with more extreme events, less number of rainy days and long dry spell in the regions where agriculture is primarily dependant on nature. The harmony in the hydrological cycle is disturbed in most of the climates over the world. Ephemeral drought has become common place even in the rainy seasons. Such drought may not be so conspicuous from the hydrologic point of view but their impact on the production system is acute. Water scarcity experienced in the critical growth stages of crops results in drastic reduction of the productivity.

Nevertheless, crop growth and yield are the result of various factors such as its genetic potential, soil characteristics, weather parameters, management practices (*viz.*, use of inputs such as fertilizers and irrigation) and biotic stress (Subhas and Ram Mohan, 2011). Of all these factors, water management is one of the principal factors affecting crop production. Further, the uncertainty in agricultural production prevails in the most developing economies due to their extra-dependence on nature. Spatial and temporal variability of rainfall is a normal phenomenon. Rainfall, though a stochastic variable, follows a certain probabilistic distribution. Due to the specific statistical behaviour of rainfall the farmers in a particular place on earth can choose their crops and can plan agricultural operations. But the present scenario of climate change is throwing all the plans hay way. Increase of non-rainy days causing serious scarcity of water in the cropping system. However, such situation may not be fit to the definition of drought for the purpose of disaster management. It is often seen that debates start among the stake holders and administrators even with the definitions of drought in order to provide relief to the affected people. This is not limited to agricultural alone. Fresh water crisis is another issue to be addressed in this changed situation. Understanding drought in historical perspectives and present situation is very much necessary now to assess and plan for fresh water utilization. Unlike flood analyses, the change in drought characteristics and the impact of climate change on it has not been explored properly (Mishra and Singh, 2010). It is beyond doubt that drought is going to be future threat to mankind hence it is pertinent to analyse the concept, definition and characterization of drought and different indices of drought characterization in the changed paradigm.

Drought is a regional phenomenon in which the natural water availability for a territorial area is lower than the normal for a significant period of time (Beran and Rodier 1985). Rossi (2000) explained the droughts as a recurring natural phenomenon caused by deficit water availability over a large geographical area for an extended period of time. The drought phenomenon are complex and many a times difficult to assess causing impacts of varying degree on nature, quality of life and economy (Hisdal and Tallaksen 2003).

Drought is a natural disaster and perhaps maximum numbers of people are affected by it. It is a creeping phenomenon and unlike other natural disaster; it does not happen suddenly. Many a times the onset of drought cannot be determined. But once it occurs, the impact of it last for a longer period affecting the economy and society of a region. The damage due to drought is non-structural *i.e.* and it affects directly on the hydrologic system leaving people longing for the elixir of life.

Droughts are normal phenomenon and their recurrences are inevitable. Due to its complexity, it is difficult to coin a universally accepted definition thereby suggesting appropriate mitigation measures (Wilhite and Glantz 1985). While defining drought, it is important to differentiate between conceptual and operational definition (Glantz 1987). Conceptual definition may be in relative term and may be more or less qualitative whereas operational definition is needed to understand the onset, severity and termination drought so that mitigation as well as necessary contingency measures can be planned ahead of such occurrences. Mishra and Singh (2010) enlisted some definitions of drought forwarded by various organisations and researchers. The World Meteorological Organisation (WMO) defines drought as sustained, extended deficiency in precipitation. The UN Convention to Combat Drought and Desertification (General U.S. 1994) defines drought as the naturally occurring phenomenon that exists when precipitation has been significantly below normal levels causing serious hydrological imbalances that adversely affect land resource production system. The Food and Agricultural Organization (FAO, 1983) defines drought hazard as the percentage of years when crops fail from lack of moisture. Schneider (1996) defines drought in encyclopaedia of climate and weather as an extended period- a season, a year or several years of deficient rainfall relative to the statistical multi-year mean for a region. Gumbel (1963) defined drought as the smallest annual value of daily stream flow. Palmer (1965) defined drought as a significant deviation from the normal hydrologic conditions of the area. Linsley *et al.* (1975) defined drought as a sustained period of time without significant rainfall.

The definitions vary, but the significantly common characteristic of these definitions is 'deficit of rainfall for a long period'. The definitions are as such more of qualitative in nature and do not specify any period of 'deficit rainfall'. Hence, the definitions warrant some quantitative classifications for further characterizations and their mitigation measures depending upon the severity and duration. Michaelides and Pashiardis (2008) suggested that drought should not be viewed solely as physical phenomenon but should be related to its impact on the society. It should be understood that incorrect estimation of drought might have serious consequences on ecology and economy due to inappropriate design of drought mitigation measures.

8.2 Classification of drought

In order to characterize the drought based on their effects on environment, hydrology and hydro-meteorology, American Society of Meteorology (2004) classified drought into four categories, *viz.* Meteorological drought, hydrological drought, agricultural drought and socio-economical drought.

Meteorological drought is defined usually on the basis of the degree of dryness (in comparison to some "normal" or average amount) and the duration of the dry period (Mishra and Desai, 2005). Precipitation is the commonly used parameter for the analysis of departure from the average to study relative dryness of the weather. Hydrological drought is concerned with the period of inadequate surface and subsurface water in order to fulfil the demand of hydrological system. The stream flow data is generally the parameters used to study the hydrological drought and their severity and impact on the water resources management systems. While mentioning about the water resources system it includes the influence of watershed characteristics and geological system prevailing in a particular area. Soil moisture availability is the yardstick of agricultural production. Inadequate soil moisture leads to crop failure. Agricultural drought refers to the deficit of soil moisture in order to support crop production processes without any reference to water resources system. Meteorological and hydrological droughts are closely intertwined with agricultural drought. Deficit of soil moisture, however, is relative to season and the stages of crops which may results in crop failure. Weather factors and biological factors are primarily influencing the water demand of a cropping system. More the demand of water for a crop in a season the more is the likelihood of water deficit in a given water scenario. On the other hand, the water deficit experienced by a water sensitive crop may not be the same to another crop. Socio-economic drought occurs when the economic good is short supply to the demand due to inadequate water availability in a water resource system leading to failure of production process. The imbalance of demand and supply is related to weather related short supply of water and not for other reasons.

Drought is a three dimensional phenomenon and is expressed in terms of severity, duration and affected area (Tigkas 2008). However estimating drought incorporating all three dimensions might be complicated and therefore many researchers have suggested for adopting one-dimensional analysis considering only the severity of the phenomenon (Tsakiris and Vangelis 2005 and Tsakiris *et al.* 2008). The analysis of severity of drought phenomenon generally provides valuable and practical results for managing its adverse impact.

Expressing drought in terms of a single indicator is always convenient for planners to take quick and appropriate decision. This single indicator can be termed as drought index indicating the severity of drought. The index can provide a clear understanding of the phenomenon over an area (Pai *et al.*, 2011). Several indices are available to express the severity of droughts. Some of the commonly used indices are given below:

8.2.1 Meteorological Drought

i. Percentage of Normal Precipitation (PNP) and Accumulated Precipitation Departure (Werick *et al.* 1994)

ii. National Rainfall Index (Gommes and Petrassi 1994)

iii. Deciles (Gibbs and Maher 1967)

iv. Standardized Precipitation Index (SPI) (McKee *et al.* 1993)

v. Reconnaissance Drought Index (RDI) (Tsakiris and Vangelis 2005)

vi. Bhalme and Mooly Drought Index (BMDI) (Bhalme and Mooley 1980)

8.2.2 Hydrological Drought

i. Surface Water Supply Index (SWSI) (Shafer and Dezman 1982)

ii. Palmer Hydrological Drought Index (PHDI) (Karl 1986)

iii. Palmer Drought Severity Index (PDSI) (Palmer 1965)

iv. Reclamation Drought Index (RDI) (Weghorst 1996)

v. Stream flow drought index

8.2.3 Agricultural Drought

i. Soil Moisture Drought Index (SMDI) (Hollinger *et al.* 1993)

ii. Crop Moisture Index (CMI) (Palmer, 1968)

iii. Crop Specific Drought Index (CSDI) (Meyer and Hubbard, 1995)

iv. Vegetation Condition Index (VCI) (Liu and Kogan, 1996) and

8.2.4 Socio-economic Drought

All the drought indices are developed based on either precipitation alone or combination of precipitation and temperature and precipitation and soil moisture. As evaporation and evapo-transpirations are the main driving forces in extraction of soil moisture, temperature plays an important role in the drought indices. The characteristics of the drought indices are given in the Table 8.1.

8.3 Meteorological Drought Indices

8.3.1 Percent Normal Precipitation (PNP)

The PNP is calculated by dividing actual rainfall by its long period average (LPA) or normal rainfall (typically considered to be a 30-year mean) and multiplying by 100%. This index is always positive with 0 as the lowest limit and there is no upper limit. The advantage of the method is its simplicity and its effectiveness when used for single region or single season.

Table 8.1: Different drought indices

Name	Factors used	Time scale	Main concept
PDSI	r, t, et, sm, rf	m	Based on moisture inflow, outflow and storage.
Deciles	r	m	Dividing the distribution of the occurrences over a long term r record into sections each representing ten percent.
CMI	r, t	w	Like the PDSI except considering available moisture in top 5 feet of soil profile.
BMDI	r	m, y	Percent departure of r from the long term mean.
SWSI	P, sn	m	Like the PDSI except considering sn.
SMDI	sm	y	Summation of daily sm for a year.
SPI	r	3m, 6m, 12m, 24m, 48m	Difference of r from the mean for a particular time and dividing it by the standard deviation.

Name	Factors used	Time scale	Main concept
PNP	r	m	Dividing actual r by the normal value
CSDI	ev	s	Summation of the value of calculated ev divided into possible ev during the growth of specific crops.
RDI	r	y, c	Patterns and abnormalities of r on a continental scale.
RAI	r	m, y	r compared to arbitrary value of +3 and -3, which is assigned to the mean of ten extreme + and - anomalies of r.
VCI	Vegetatio n Index	-	Based on satellite data

Abbreviations:

P: factors used in PDSI, r: precipitation, et: evapotranspiration, ev: evaporation, t: temperature, sm: soil moisture, rf: runoff, sn: snow pack, w: week, m: month, s: season, y: year, c: century, 3 m: 3 months, etc.

$$PNPI = \frac{Pi}{P} \times 100 \qquad \text{...... (8.1)}$$

Where P_i is year I rainfall and P is mean rainfall in statistical years. Values less than 40 are categorised as drought.

However, there are some disadvantages associated with it. One of them is that that the mean, or average, precipitation is often not the same as the median precipitation, which is the value exceeded by 50% of the precipitation occurrences in a long-term climate record. The reason for this is that precipitation on monthly or seasonal scales does not have a normaldistribution. Use of the percent of normal comparison implies a normal distribution where the mean and median are considered being the same. Because of the variety in the precipitation records over time and location, there is no way to determine the frequency of the departures from normal or compare different locations. This makes it difficult to link a value of a departure with a specific impact occurring as a result of the departure, inhibiting attempts to mitigate the risks of drought based on the departures from normal and form a plan of response (Willeke *et al.* 1994).

The seasonal rainfall is classified as moderate meteorological drought, when it is 50–74% of LPA and a severe meteorological drought when it is less than 50% of LPA.

8.3.2 National Rainfall Index

A national rainfall index (NRI) is defined for areas where water is limiting for agricultural production. The NRI can be used as an indicator of the quality of the agricultural season. The definition as a national annual value allows a comparison of countries and years and allows a comparison with other country wide statistical indicators. The NRI is defined as the national average of the total annual precipitation weighted by its long-term average.

8.3.3 Deciles

Arranging monthly precipitation data into deciles is another drought-monitoring technique. It was developed by Gibbs and Maher (1967) to avoid some of the weaknesses within the "percent of normal" approach. The technique they developed divided the distribution of occurrences over a long-term precipitation record into tenths of the distribution. They called each of these categories a *decile*. The first decile is the rainfall amount not exceeded by the lowest 10% of the precipitation occurrences. The second decile is the precipitation amount not exceeded by the lowest 20% of occurrences. These deciles continue until the rainfall amount identified by the tenth decile is the largest precipitation amount within the long-term record. By definition, the fifth decile is the median, and it is the precipitation amount not exceeded by 50% of the occurrences over the period of record. The deciles are grouped into five classifications (Table 8.2).

In this approach suggested by Gibbs and Maher (1967), monthly precipitation totals from a long-term record are first ranked from highest to lowest to construct a cumulative frequency distribution. The distribution is then split into 10 parts (tenths of distribution or deciles). The first decile is the precipitation value not exceeded by the lowest 10% of all precipitation values in a record. The second decile is between the lowest 10 and 20% etc. Comparing the amount of precipitation in a month (or during a period of several months) with the long-term cumulative distribution of precipitation amounts in that period, the severity of drought can be assessed. The deciles are grouped into five classes, two deciles per class. If precipitation falls into the lowest 20% (deciles 1 and 2), it is classified as much below normal. Deciles 3 to 4 (20 to 40%) indicate below normal precipitation, deciles 5 to 6 (40 to 60%) indicate near normal precipitation, 7 and 8 (60 to 80%) indicate above normal precipitation and 9 and 10 (80 to 100%) indicate much above normal precipitation. In the current study, monthly rainfall time series are normalized using the Box–Cox transformation (McMahon and Mein 1986).

Table 8.2: Classification drought severity based on deciles values

Deciles class	Category of drought
deciles 1-2: lowest 20%	much below normal

Deciles class	Category of drought
deciles 3-4: next lowest 20%	below normal
deciles 5-6: middle 20%	near normal
deciles 7-8: next highest 20%	above normal
deciles 9-10: highest 20%	much above normal

The deciles method was selected as the meteorological measurement of drought within the Australian Drought Watch System because it is relatively simple to calculate and requires less data and fewer assumptions than the Palmer Drought Severity Index (Smith *et al.*, 1993). In this system, farmers and ranchers can only request government assistance if the drought is shown to be an event that occurs only once in 20–25 years (deciles 1 and 2 over a 100-year record) and has lasted longer than 12 months (White and O'Meagher 1995). This uniformity in drought classifications, unlike a system based on the percent of normal precipitation, has assisted Australian authorities in determining appropriate drought responses. One disadvantage of the decile system is that a long climatological record is needed to calculate the deciles accurately.

8.3.4 Standardized Precipitation Index (McKee *et al.* 1993)

Standardized Precipitation Index (SPI) is a simple and index which is calculated from the long term record of precipitation in each location (at least 30 years) for a desired period. This long-term record is fitted to a probability distribution, which is then transformed to a normal distribution so that the mean SPI for the location and desired period is zero (McKee *et al.* 1993; Edwards and McKee 1997). SPI is used to identify the meteorological drought or deficit of precipitation.

The SPI is classified into 5 categories as shown in Table 8.3. The index is effective in providing early warning of drought and its severity because it can specify for each location and is well-suited for risk management. Since SPI takes into account the stochastic nature of the drought and is therefore a good measure of short- and long-term meteorological drought. However, for a longer time scale say 12-24 months period the data can change and becomes less reliable. SPI does not account for the effect of soil, land use characteristic, crop growth, and temperature anomalies that are critical for agricultural drought monitoring.

The SPI is computed by fitting a probability density function to the frequency distribution of precipitation summed over the time scale of interest. This is performed separately for each month (or any other temporal basis of the raw precipitation time series) and for each location in space. Each probability density function is then transformed into a standardized normal distribution. The gamma distribution is defined by its probability density function is:

$$g(x) = \frac{1}{\beta^{\alpha}\Gamma(\alpha)} x^{\alpha-1} e^{-x/\beta} \text{ for } x > 0 \qquad \text{...... (8.2)}$$

Where, α>0 is a shape factor, β >0 is a scale factor, and x >0 is the amount of precipitation. $\Gamma(\alpha)$ is the gamma function which is defined as:

$$\Gamma(\alpha) = \int_{0}^{\infty} y^{\alpha-1} e^{-y} dy \qquad \text{...... (8.3)}$$

Fitting the distribution to the data requires that α and β be estimated. Edwards and McKee (1997) suggested a method for estimating these parameters using the approximation of Thom (1958) for maximum likelihood as follows:

$$\overline{\alpha} = \frac{1}{4A}\left(1 + \sqrt{1 + \frac{4A}{3}}\right) \qquad \text{...... (8.4)}$$

$$\overline{\beta} = \frac{\overline{x}}{\hat{\alpha}} \qquad \text{...... (8.5)}$$

Whare $A = \ln \overline{x} - \frac{\sum_{i=1}^{n} \ln x}{n}$ for n observation

Table 8.3: Classification of SPI

≥ 2.0	extremely wet
1.5 -1.99	very wet
1.0 - 1.49	moderately wet
-0.99 - 0.99	near normal
-1.0 - -1.49	moderately dry
-1.5 - -1.99	severely dry
≤ -2.0	extremely dry

8.3.5 Reconnaissance Drought Index (RDI)

Reconnaissance drought Index is a new drought and is being successfully applied in number of Mediterranean countries (Tsakiris and Vangelis 2005 and Tsakiris 2007). It is a ratio of aggregated quantities of precipitation and potential evapotranspiration. Although it behaves similar to SPI but its provides better understanding as it takes both precipitation and potential evapo-transpiration (PET) into consideration which are directly affected by climate change. Further it provides scope for rational comparison of drought between areas of different climatic characteristics. The *RDI* is expressed as:

$$\alpha_k^{(i)} = \frac{\sum_{j=1}^{k} P_{ij}}{\sum_{j=1}^{k} PET_{ij}}, \ i = 1 \ to \ N \qquad \text{...... (8.6)}$$

Where P_{ij} and PET^{ij} are the precipitation and the potential evapotranspiration of month *j* of year *i*, starting usually from October and *N* is the total number of years of the available data.

The RDI is also expressed as standardized RDI and is computed using the following equation:

$$RDI_{std} = (K) = \frac{y_k = \bar{y}_k}{\bar{\sigma}_k} \qquad \text{...... (8.7)}$$

Where y_k is ln a_k, $\bar{y}_k$ is the arithmetic mean and $\bar{\sigma}_k$ is the standard deviation. The RDI_{std} is useful when comparisons are made between relative standing of items from distributions with different means and/or different standard deviations. The RDI_{std} is similar to *SPI* which is a special application of transformation rules. The transformation of *RDI* into RDI_{std} will necessarily have a mean of zero and standard deviation of one (Sirdas and Sen 2003). The RDI_{std} values can be compared with the same threshold as the *SPI* (Table 8.3)

Earlier researchers have advised that RDI is computed on aggregate time scales viz., 3,6,9 and 12 months (Tsakiris and Vangelis 2005).

Earlier researchers have advised that *RDI* should be computed on aggregate time scales viz., 3, 6, 9 and 12 months (Tsakiris and Vangelis 2005).

Advantages of RDI:

— Reconnaissance drought index is physically based, since it calculates the aggregated deficit between the evaporative demand of the atmosphere and precipitation.

— It can be calculated for any period of time, however, 3, 6, 9 and 12 months' periods are proposed by the authors.

— It can be more effectively associated with agricultural and hydrological drought.

— The annual value of initial index a12 may be directly compared with the aridity index of the area.

— It is considered to be an ideal index to study the effects of climate instability conditions.

8.3.6 Bhalme and Mooly Drought Index (BMDI)

This index was developed by Bhalme & Mooley (1980) and was based on the four months of monsoon rainfall in India. To use this type of index, a moisture index is first defined as the percentage of departure of monthly rainfall from the long term mean weighted by the reciprocal of the coefficient of variation and then an appropriate drought index derived from the moisture index to account for the duration factor of abnormal moisture deficiency with the monthly index values generally ranging from -4 to +4 in which the descriptive terms used to characterise the weather being similar to those of PDSI.

8.4 Hydrological Drought Indices

8.4.1 Surface Water Supply Index (SWSI)

The objective of the SWSI was to incorporate both hydrological and climatological features into a single index value resembling the Palmer Index for each major river basin where local precipitation is not the only source of water (Shafer and Dezman 1982). The SWSI is calculated based on monthly non-exceedance probability from available historical records of reservoir storage, stream flow, snow pack, and precipitation. Because it is dependent on the season, the SWSI is computed with only snowpack, precipitation, and reservoir storage in the winter. During the summer months, stream flow replaces snowpack as a component within the SWSI equation.

The purpose of SWSI is primarily to monitor the abnormalities in surface water supply sources. Hence it is a good measure to monitor the impact of hydrologic drought on urban and industrial water supplies, irrigation and hydroelectric power generation. However, there is a time lag before precipitation deficiencies are detected in surface and subsurface water sources. As a result, the hydrological drought is out of phase from agricultural drought. Because of this phase difference, SWSI is not a suitable indicator for agricultural drought (Narasimhan and Srinivasan 2005).

One of the limitations of the index is its location specific nature. The index of different basins cannot be compared with each other.

8.4.2 Palmer hydrological drought index (PHDI) (Karl 1986)

PHDI uses a modification of the PDSI to assess moisture anomalies that affect stream flow, ground water, water storage, being more sensitive to hydrological components

- PHDI has more stringent criterion for the elimination of the drought or wet spells, which results in index rebounding gradually (more slowly than PDSI) toward the normal state.

- PDSI considers drought finished when a moisture conditions begin terminated an uninterrupted rise that ultimately erases water deficit; PHDI considers drought ended when the moisture deficit actually vanishes PHDI has been derived from the PDSI index to quantify the long term impact from hydrological drought.

Pros: The PHDI has been officially used by NCDC to determine the precipitation needed for drought termination and amelioration which has a PHDI equal to -0.5 and -2.0, consecutively. It has been used in Indiana for drought monitoring.

Cons: The PHDI is developed from precipitation, outflow, and storage. PHDI may change more slowly than PDSI and it has sluggish response for drought.

8.4.3 Palmer Drought Severity Index (PDSI)

Palmer Drought Severity Index (PDSI) is a meteorological drought index developed by Wayne Palmer in 1965. This was the most comprehensive effort to assess the total moisture status of a region. The index was developed to measure the departure of the moisture supply. Palmer based his index on the supply-and demand concept of the water balance equation, taking into account more than just the precipitation deficit at specific locations. It is calculated based on precipitation and temperature data, as well as the local Available Water Content (AWC) of the soil. From the inputs, all the basic terms of the water balance equation can be determined, including evapotranspiration, soil recharge, runoff, and moisture loss from the surface layer. Running the PDSI requires temperature, precipitation, and soils information. The PDSI values for classification drought severity are given in Table 8.4.

Table 8.4: Classification of PDSI based on PDSI values

PDSI Values	Classification
4.0 or more	Extremely wet
3.0 to 3.99	Very wet
2.0 to 2.99	Moderately wet
1.0 to 1.99	Slightly wet
0.50 to 0.99	Incipient wet spell
0.49 to -0.49	Near normal
-0.99 to -0.50	Incipient Dry spell
-1.0 to -1.99	Mild drought
-2.0 to -2.99	Moderate drought
-3.0 to -3.99	Severe drought
-4.0 or less	Extreme drought

PDSI is perhaps the most widely used regional drought index for monitoring droughts. The index has been used to illustrate the areal extent and severity of various drought episodes (Palmer 1968; Karl and Quayle 1981) and to investigate the spatial and temporal drought characteristics (Lawson *et al.* 1971; Klugman1978; Karl and Koscielny 1982; Diaz 1983; Jones *et al.* 1996) as well as to explore the periodic behavior of droughts (Rao and Padmanabhan 1984), monitoring hydrologic trends, crop forecasts, and assessing potential fire severity (Heddinghaus and Sahol 1991), droughts over large geographic areas (Johnson and Kohne, 1993), and drought forecasting (Kim and Valdes 2003; Özger *et al.* 2009).

Limitations of PDSI include: (1) an inherent time scale making PDSI more suitable for agricultural impacts and not so much for hydrologic droughts, (2) assumptions that all precipitation is rain, thus making values during winter months and at high elevations often questionable. PDSI also assumes that runoff only occurs after all soil layers have become saturated, leading to an underestimation of runoff, and (3) PDSI can be slow to respond to developing and diminishing droughts (Hayes *et al.*, 1999).The PDSI does not consider human impacts on the water balance, such as irrigation. Palmer developed criteria for determining when a drought or a wet spell begins and ends, which adjust the PDSI accordingly.

In near real time, Palmer's index is no longer a meteorological index but becomes a hydrological index referred to as the Palmer Hydrological Drought Index (PHDI) because it is based on moisture inflow (precipitation), outflow, and storage, and does not take into account the long term trend (Karl and Knight, 1985). PDSI is designed for agriculture and does not accurately represent the hydrological impacts resulting from longer droughts. The PDSI does not do well in regions where there are extremes in the variability of rainfall or runoff. The "extreme" and "severe" classifications of drought occur with a greater frequency in some parts of the country than in others (US). "Extreme" droughts in the Great Plains occur with a frequency greater than 10%. This limits the accuracy of comparing the intensity of droughts between two regions and makes planning response actions based on certain intensity more difficult.

The values quantifying the intensity of drought and signalling the beginning and end of a drought or wet spell were arbitrarily selected based on Palmer's study of central Iowa and western Kansas and have little scientific meaning. The Palmer Index is sensitive to the available water content (AWC) of a soil type. Thus, applying the index for a climate division may be too general. The two soil layers within the water balance computations are simplified and may not be accurately representative of a location. Snowfall, snow cover, and frozen ground are not included in the index. All precipitation is treated as rain, so that the timing of PDSI or PHDI values may be inaccurate in the winter and spring months in regions where snow occurs. The natural lag between when precipitation falls and the resulting runoff is not considered. In addition, no runoff is allowed to take place in the model until the water capacity of the surface and subsurface soil layers is full, leading to an underestimation of runoff. Potential evapotranspiration is estimated using the Thornthwaite method. This technique has wide acceptance, but it is still only an approximation.

8.4.4 Reclamation Drought Index (RDI) (Weghorst, 1996)

The reclamation drought, RDI just like the SWSI, is calculated at the river basin level, and makes use of temperature, precipitation, snow pack, stream flow and reservoir levels as data input and has the advantage that by including a temperature component, it also accounts for evaporation. The index was developed by Bureau of Reclamation in the USA, as a trigger to release drought emergency relief funds and has recently been used as a tool to define drought severity and duration, and to predict the onset and end of drought periods.

The index is limited in its use for inter basin comparison as it is unique to each river. The classification of the index are given in Table 8.5. Similar to the SWSI, the RDI is calculated at river basin level, and incorporates the supply components of precipitation, snow-packs, stream flow and reservoir levels. However, it differs from SWSI with RDI builds a temperature based demand component and duration component into the index. The RDI is adaptable to each particular region and its main strength lies in its ability to account for both climate and water supply factors. The RDI values and severity designations are similar to SPI, PDSI and SWSI.

Table 8.5: Reclamation Drought Index Classification Criteria

Value	Condition
≥ 4.0	Extremely wet
1.5 - 4.0	Moderately wet
1.0 - 1.5	Normal to mild wetness
0.0 - -1.5	Normal to mild drought
-1.5 - -4.0	Moderate drought
≤ -4.0	Extremely drought

8.4.5 Stream flow drought index

Stream flow drought Index (SDI) is a simple and effective index. It is based on cumulative flow volume for overlapping period of three six, nine and twelve months within each hydrological period. It allows defining drought states which are modelled as a non stationary (Nalbantis 2008).

It is assumed that a time series of monthly stream flow volumes $Q_{i,j}$ are available where 'i' denotes the hydrological year and the j_{th} month within that hydrological year ($j=1$ for October and $j=12$ for September). Based on this series we obtain:

$$Vi,k = \sum_{j=1}^{3k} Q_{i,j};\ i=1,2\ldots\ldots \text{ and } j=1,2,\ldots\ldots;\ k=1,2,3,4 \quad \ldots\ldots (8.8)$$

Where, $V_{i,k}$ is the cumulative stream flow volume for the ith hydrological year and k^{th} reference period , k=1 for October to December; k= 2 for October to March; k=3 for October to June and k=4 for October to September.

Based on the cumulative stream flow volume $V_{i,k}$ the stream flow drought index (SDI) is defined for each reference period k of the ith hydrological year as follows:

$$SDI_{i,k} = \frac{V_{i,k} - \bar{V}_k}{S_k};\ i = 1,2 \ldots \text{ and } k = 1,2,3,4 \quad \ldots\ldots (8.9)$$

Where, $\bar{V}_k$ and S_k are respectively the mean and standard deviation of the stream flow volumes of a reference period k as these are estimated over a long period of time. The streamflow drought index (SDI) is identical to SPI and RDI.

8.5 Agricultural Drought Indices

8.5.1 Crop moisture index (CMI) (Palmer 1968)

Palmer (1968) developed the CMI as an index for short-term agricultural drought from procedures within the calculation of the PDSI. CMI is most effective for measuring agricultural droughts during warm seasons (i.e., growing season) (Heim 2002). The CMI reflects moisture supply in the short term across major crop-producing regions and is not intended to assess long-term droughts. PDSI is calculated from precipitation deficits for monitoring long-term drought conditions, whereas CMI is calculated from evapo-transpiration deficits for monitoring short-term agricultural drought conditions that affect crop growth. The CMI responds rapidly to changing conditions may provide misleading information about long-term conditions.

Limitation of CMI is that with increase in potential evapo-transpiration the CMI values also increases. An increased value of CMI indicates wetter moisture conditions there is no case in nature where an increase in potential evaporation would produce wetter moisture conditions and such a case was reported by Juhasz and Kornfield (1978). The unnatural response of CMI to changes in temperature is due to the dependence of the abnormal evapo-transpiration term on the magnitude of potential evapo-transpiration.

8.5.2 Soil Moisture Drought Index (SMDI) (Hollinger *et al.* 1993)

Soil Moisture Deficit Index (SMDI) is computed as the weekly soil moisture normalized by long-term statistics. Weekly values are then added on an incremental basis to account for the duration of a drought. The SMDI is computed separately for different soil depths in order to consider the varying rooting depth for different crops and stages of plant development. The Evapo-Transpiration Deficit Index (ETDI) is computed similar to the SMDI, but considers the water stress ratio of potential to actual evapo-transpiration instead of soil moisture

Compared to PDSI and SPI, the two new indices SMDI and ETDI show in general a high correlation, indicating a satisfying identification of drought events. In detail, ETDI and SMDI for low soil depths were better correlated to short-term SPI (1-3 months), while the SMDI for deeper soil layers corresponded better to the more slowly reacting PDSI and SPI of 6-12 months.

8.5.3 Crop Specific Drought Index (CSDI) (Meyer and Hubbard 1995)

During studies it was observed that each crops responds differently to drought so the indices such as CMI or PDSI did not reflect the impact. Both the indices were not reliable indices. So Meyer *et al.*, (1993) proposed Crop Specific Drought Index (CDSI). The index was developed to examine the actual impact of drought on individual crop. The methodology used for estimation of the drought is:

$$CSDI_{act} = \frac{Y_{act}}{Y_{pot}} \qquad \text{...... (8.10)}$$

$$CSDI_{pred} = \frac{Y_{pred}}{Y_{pot}} = \prod_{i=1}^{n} \left(\frac{\sum ET_{cal}}{\sum ET_{pot}} \right)_i^{\lambda_i} \qquad \text{...... (8.11)}$$

Where, Y_{act}, Y_{pred} and Y_{pot}, are the actual yield(kg ha^{-1}), predicted (kg ha^{-1}) and potential yield (i.e., yield when soil moisture is not limiting) (kg ha^{-1}), respectively; ET_{calc} and ET_{pot} are the estimated evapo-transpiration and potential evapo-transpiration when soil moisture is not limiting, respectively; n represents the number of growth periods that comprise the crop's growth cycle, and λ_i is the relative sensitivity of the crop to moisture stress during the i '~ period of growth.

The inputs required for developing the index includes climatic parameters viz., daily maximum and minimum temperature, rainfall, dew point temperature, wind speed and global solar radiation. Soil profile characteristics, yield and phenology data are required for proper correlations to growing days, crop progress and final yield. The strengths and weakness of the method are as follows:

Strengths: Very specific to a particular crop and based upon the development of the plant. The model takes into account when the drought stress occurred during plant growth and estimates the overall impact on yield.

Weaknesses: The inputs are quite complex and many locations will lack the required instruments or period of record needed to properly assess conditions.

8.5.4 Vegetation Condition Index (VCI) (Liu and Kogan 1996)

Since the 1970s, several studies have used satellite land observation data to monitor a variety of dynamic land surface processes (e.g., Reed *et al.* 1994; Yang *et al.* 1998; Peters *et al.* 2002; Gu *et al.* 2007). Satellite remote sensing provides a synoptic view of the land and a spatial context for measuring drought impacts which have proved to be a valuable source of timely, spatially continuous data with improved information on monitoring vegetation dynamics over large areas. The vegetation condition index (VCI), computed from satellite advanced very high resolution radiometer (AVHRR) radiance (visible and near infrared) data adjusted for land, climate, ecology, and weather conditions, shows promise when used for drought detection and tracking (Kogan 1995). The VCI allows detection of drought and measurement of the time of its onset and its intensity, duration, and impact on vegetation. However, since the VCI is based on vegetation, it is primarily useful for the summer growing season. It has limited utility for cold seasons when vegetation is largely dormant (Heim, 2002).

Vegetative and agricultural droughts reflect vegetation stress. NDVI reflects the vegetation condition through the ratio of responses in near infrared (Ch2) and visible (Ch1) bands of Advanced Very High Resolution Radiometer (AVHRR) of NOAA. It is expressed as:

$$\text{NDVI} = (\text{Ch2} - \text{Ch1}) / (\text{Ch2} + \text{Ch1}) \quad \text{...... (8.12)}$$

VCI, TCI, and VHI have been developed further using the following equations as:

$$\text{VCI} = 100^*(\text{NDVI} - \text{NDVI}_{min}) / (\text{NDVI}_{max} - \text{NDVI}_{min}) \quad \text{...... (8.13)}$$

$$\text{TCI} = 100^*(\text{BT}_{max} - \text{BT}) / (\text{BT}_{max} - \text{BT}_{min}) \quad \text{...... (8.14)}$$

$$\text{VHI} = 0.5(\text{VCI}) + 0.5(\text{TCI}) \quad \text{...... (8.15)}$$

where, NDVI, NDVI_{min}, and NDVI_{max} are the seasonal average of smoothed weekly NDVI, its multi-year absolute minimum and its maximum respectively; BT, BTmin, and BTmax are similar values for brightness temperature (Kogan, 2001). Brightness temperature values are obtained from the thermal band (Ch4) of NOAA-AVHRR. NDVI generally provides abroad overview of the vegetation condition and spatial vegetation distribution in a region.

Vegetative drought is closely related with weather impacts. However, in NDVI, the weather component gets subdued by strong ecological component. VCI separates the short-term weather-related NDVI fluctuations from the long-term ecosystem changes (Kogan 1990, 1995). Therefore, while NDVI shows seasonal vegetation dynamics, VCI rescales vegetation dynamics in between 0 and 100 to reflect relative changes in the vegetation condition from extremely bad to optimal (Kogan 1995, 2003). VCI and TCI characterises respectively the moisture condition and thermal condition of vegetation while VHI represents overall vegetation health (Kogan and Zhu, 2001). Since favourable weather provides optimal moisture condition, high values of VCI correspond to healthy and unstressed vegetation. On the other hand, low TCI values correspond to vegetation stress due to dryness by high temperature. TCI provides opportunity to identify subtle changes in vegetation health due to thermal effect as drought proliferates if moisture shortage is accompanied by high temperature (Kogan 2002). During calculation of VHI, an equal weight has been assumed and assigned to both VCI and TCI since moisture and temperature contribution during a vegetation cycle is currently not known (Kogan 2001). Regional VHI maps have been generated through interpolation of the point values and classified using the scheme developed by Kogan.

8.6 Socio-Economic Drought Index

8.6.1 Multivariate Standardized Reliability and Resilience Index (MSRRI) (Mehran *et al.* 2015)

The Multivariate Standardized Reliability and Resilience Index (MSRRI) was proposed by Mehran *et al.* 2015). The index is based on type of reservoir system on a selected time frame (within year or over year). After the time frame is defined two more indicators are defined viz., water storage resilience indicator and inflow - demand reliability indicator. Inflow - demand reliability (IDR) indicator is derived by computing the sum of the per cent change of inflow with respect to water demand during the projected time frame:

$$\alpha_t = \frac{\sum_{i=t-m+1}^{t} Q_{in_t} - Q_{\text{est}_t}}{Q_{\text{est}_t}}, \; Q_{\text{est}_t} = \begin{cases} \sum_{i=t-12}^{t-13+m} (Q_{\text{out}})i, & \text{if } m = 6 \\ \sum_{i=t-m+1}^{t} (Q_{\text{out}})i, & \text{if } m = 12 \end{cases} \quad ......(8.16)$$

where, Q_{in_t} is the monthly inflow to the reservoir (i ranges from month 1 to N, which is the sample size), m is the selected time frame in months (6 months for within - year and 12 for over - year systems), Q_{est_t} is the total estimated water demand during projected time frame, and time step t ranges from month 13 to N ($t = 13, N$). The total water demand for the projected time frame (next m months) is estimated based on the same period in the previous year. For this reason, the index can only be estimated starting from the second year of the data ($t = 1,..., 12$).

The other indicator introduced here in respect to "bottom - up" methodology is the water storage resilience indicator. This indicator is defined based on monthly inflow, monthly water demand, monthly storage, and total water demand during the time frame. Water storage resilience (WSR) indicator is computed on a monthly basis and shows whether the reservoir storage is sufficient to satisfy water demand for the selected time period (m):

$$\beta_t = \frac{\mathrm{St}_t + Q\mathrm{in}_t - Q\mathrm{out}_t = -O_{\min} - Q\mathrm{est}_t}{Q\mathrm{est}_t} \qquad \text{...... (8.17)}$$

where St_t is the reservoir storage at month t, where $t = 13,..., N$, $O_{\min}$ is the reservoir minimum operational storage, Qin_t is the monthly inflow to the reservoir at month t, $Qest_t$ is the total estimated water demand during projected time frame (either 6 or 12 months) as discussed before, and $Qout_t$ is the monthly water demand at month t. If reservoir storage is not available, then a reservoir model is needed to estimate the storage based on the inflow and outflow (demand).

For convenience and easy cross comparison, the two indicators are standardized using the standard normal distribution. First, the marginal probabilities of both indicators (water storage resilience and inflow - demand reliability) are calculated (Gringorten 1963).

$$P(X_t) = \frac{l - 0.44}{N + 0.12} \qquad \text{...... (8.18)}$$

Where, N is the sample size, l denotes the rank of nonzero indicator (α or β) data from the smallest to largest, and $P(x_t)$ is the corresponding empirical probability at month t. The empirical probabilities are then transformed into a standardized index (SI) as:

$$SI(x) = \varphi^{-1}(P(x)) \qquad \text{...... (8.19)}$$

$$SI(P(x)) = \begin{cases} \text{if } 0 < P(x) \le 0.5, \; + \left(k - \dfrac{C_0 + C_1 k + C_2 k^2}{1 + d_1 k + d_2 k^2 + d_3 k^3} \right) \text{ and } k = \sqrt{ln\left(\dfrac{1}{P(x)^2} \right)} \\ \text{if } 0 < P(x) \le 1, \; - \left(k - \dfrac{C_0 + C_1 k + C_2 k^2}{1 + d_1 k + d_2 k^2 + d_3 k^3} \right) \text{ and } k = \sqrt{ln\left[\dfrac{1}{(1 - P(x))^2} \right]} \end{cases}$$

...... (8.20)

where, φ is the standard normal distribution function and $P(x)$ is the empirical probability. One can also standardize the empirical probability values by a commonly used approximation in which

$$C_0 = 2.515517,\ C_1 = 0.802583,\ C_2 = 0.010328,$$

$$d_1 = 1.432788,\ d_2 = 0.189269,\ \text{and}\ d_3 = 0.$$

Substituting α and β with x from above equations leads to standardized indices for inflow - demand and water storage resilience (hereafter $SI(\alpha)$ and $SI(\beta)$). The two univariate indicators are then combined using a multivariate framework

$$Pj_t = \Pr\left(SI(\alpha)\right) \leq SI(\alpha_t),\ SI(\beta) \leq SI(\beta_t) \qquad \text{...... (8.21)}$$

where, P_{jt} is the joint (multivariate) empirical probability at month t, calculated using the two indices: inflow - demand reliability index $SI(\alpha_t)$ and water storage resilience index $SI(\beta_t)$. Having the two univariate indicators, the joint empirical probability can be derived using the multivariate model of the Gringorten plotting position introduced by Yue *et al.* 1999:

$$Pj_t(SI(\alpha_t),\ SI(\beta_t)) = \frac{l - 0.44}{N + 0.12} \qquad \text{...... (8.22)}$$

where, l is the number of occurrences of the pair $(SI(\alpha_t), SI(\beta_t))$ for $SI(\alpha) \leq SI(\alpha_t)$ and $SI(\beta) \leq SI(\beta_t)$, and N is the sample size. We define the Multivariate Standardized Reliability and Resilience Index (MSRRI) by standardizing the joint distribution function of the inflow demand reliability index and water storage resilience index (Hao and Agha Kouchak 2014):

$$\text{MSRRI} = \varphi^{-1}(P_j) \qquad \text{...... (8.23)}$$

Where, the joint empirical probability P_j can be standardized using equation. The final index (MSRRI) is not only a hybrid index (consisting of two fundamentally different indicators) but also covers both common approaches in vulnerability studies (top - down and bottom-up). MSRRI can be considered as a measure of socioeconomic drought since it evaluates the supply and storage relative to the demand. Similar to other standardized drought indicators, positive values indicate sufficient water to satisfy demand, while negative values indicate shortage of water relative to the demand.

8.7 Drought prediction and forecasting

Once the estimation of drought intensity is accomplished, its prediction and forecasting becomes important for taking up appropriate drought management strategies. Number of frequency distribution models has been adopted by various researchers. The purpose of the frequency distribution is to monitor and manage the hydrological activities like drought, floods, and other such events. It is therefore necessary to develop a mathematical formula for fitting the empirical series. If the theoretical distribution fits the empirical distribution reasonably well, it can be used to convert the confidence limits of frequencies or return periods into the confidence limits of the hydrologic phenomenon studied and it can be interpolated and extrapolated to predict extreme future event.Although number of frequency distribution are available but applicability of appropriate function is still unknown (Table 8.6). The earlier researches have reported that drought events generally follows Lognormal, Gamma, Extreme value- type I; Weibull distribution (Mishra and Desai 2005; Tsakiris and Vangelis 2005; Tigkas 2008; Yusof and Hui-Mean 2012). Su *et al.* (2009) reported the use of multi parameter distribution for statistical modelling of extreme events such as precipitation as among the recent developments. Wakeby distribution is also attempted which takes five parameters into consideration. This distribution is more flexible than the other conventional distributions and its advantage lies in the fact that it separates the distribution's extreme tails (Arbeláez *et al.* 2007). Capability to handle the separation effect better than conventional distributions makes it easier for use in many applications even with limited records (viz. 10, 20 and 30 years) (Houghton 1978). Apart from Wakeby distribution three parameter Lognormal, Gamma,Weibull distributions are also normally used.

Table 8.6: Description of frequency distribution and their parameters

Sl No	Name	PDF	Parameters
1	Lognormal (3P)	$f(x)=\dfrac{\exp\left(-\frac{1}{2}\left(\frac{\ln(x-\gamma)-\mu}{\sigma}\right)^2\right)}{(x-\gamma)\sigma\sqrt{2\pi}}$	σ- continuous parameter (σ>0) μ- continuous parameter γ- continuous location parameter (γ≡0 yields two parameter lognormal distribution)
2	Lognormal (2P)	$f(x)=\dfrac{\exp\left(-\frac{1}{2}\left(\frac{\ln(x-\mu)}{\sigma}\right)^2\right)}{(x)\sigma\sqrt{2\pi}}$	
3	Gamma (3P)	$f(x)=\dfrac{(x-\gamma)}{\beta^{\alpha}\Gamma(\alpha)}\exp\left(-\dfrac{(x-\gamma)}{\beta}\right)$	Where Γ is the gamma function and where α > 0 is a

Sl No	Name	PDF	Parameters
4	Gamma (2P)	$f(x)=\dfrac{x^{\alpha-1}}{\beta^{\alpha}\Gamma(\alpha)}\exp\left(-\dfrac{x}{\beta}\right)$	shape factor, β > 0 is a scale factor, and x > 0 is the amount of any variable
5	Weibull (3P)	$f(x)=\dfrac{\alpha}{\beta}\left(\dfrac{x-\gamma}{\beta}\right)^{\alpha-1}\exp\left(-\left(\dfrac{x-\gamma}{\beta}\right)^{\alpha}\right)$	α- continuous shape parameter (α>0) β- continuous scale parameter (β > 0)
	Weibull (2P)	$f(x)=\dfrac{\alpha}{\beta}\left(\dfrac{x}{\beta}\right)^{\alpha-1}\exp\left(-\left(\dfrac{x}{\beta}\right)^{\alpha}\right)$	γ- continuous location parameter (γ≡ 0 yields the two parameter Weibull distribution)
6	General Extreme Value-Type I	$f(x)=\dfrac{1}{\sigma}\exp\left(-(1+\kappa z)^{1/\kappa}\right)$ $(1+\kappa z)^{-1-1/\kappa}$; for $\kappa=0$ $f(x)=\dfrac{1}{\sigma}\exp(-z-\exp(-z))f$ or $\kappa=0$	κ- continuous shape parameter σ- continuous scale parameters (σ >0) μ- Continuous location parameters
7	Wakeby	$x(F)=\xi+\dfrac{\alpha}{\beta}\left(1-(1-F)^{\beta}\right)-$ $\dfrac{\gamma}{\delta}\left(1-(1-F)^{\delta}\right)$	α ≠ 0 or γ ≠ 0 β + δ > 0 or β = γ= δ =0 If α = 0 then β =0 If γ =0, then δ =0 γ ≥ 0 and α +γ ≥ 0

8.7.1 Goodness of fit

Once the frequency distribution model is selected, it is necessary to be tested for Goodness of fit. Goodness of fit tests describes the fitness of a distribution to a set of observations. Measures of Goodness of fit summarize the discrepancy between observed values with the expected values under the specific model selected. The best fitted distribution is determined based on the minimum error produced, which is measured by the acceptance and rejection of hypothesis. Some of the commonly used Goodness of Fit criteria are shown in Table 8.6.

Table 8.7: Goodness of fit criteria

Sl No	Criteria	Description	The hypothesis regarding the distributional form is rejected at the chosen significance level (*alpha*)
1	Kolmogorov-Smirnov Test	$D_n = \sup_x \lvert F_n(x) - F(x) \rvert$	If the test statistic, D, is greater than the critical value obtained from a table.
2	Anderson-Darling Test	$A^2 = -n - \frac{1}{n}\sum_{i=1}^{n}(2i-1)$ $\left[lnF(X_i) + ln(1 - F(X_{n-i+1})) \right]$	If the test statistic, A^2, is greater than the critical value obtained from a table
3	Chi-Squared Test	$\chi^2 = \sum_{i=1}^{k} \frac{(O_i - E_i)^2}{E_i}$ Where O_i is the observed frequency and Ei is the expected frequency	if the test statistic is greater than the critical value defined as $\chi^2_{1-\alpha, k-1}$

8.7.2 Return period

Estimation of return period in hydrological analysis is a important parameter for design of suitable approach for mitigating extreme hydrological event. The return period is the frequency with which on an average a given hydrological event (in the present case drought) is equalled or exceeded at least one time in this period. It can be expressed as:

$$\left(P(X \geq xT \right) = \frac{1}{T_R} \quad \text{...... (8.24)}$$

Where, $P(X \geq xT)$ is the frequency of exceedence and T_R is the return period of the event.

8.7.3 Confidence limit

As the hydrological events are highly uncertain, it is always recommended to predict within certain confidence limits. The confidence limits increase the reliability of the prediction. Oosterbaan (1994) suggested methodology for determination of the confidence interval of probability function using Student's t-statistic (t). The study stated at 90% confidence limits the t-value is close to 1.7 when N>10. The upper (U) and lower limit (L) is estimated by the following relation:

$$U = F(x) + 2 \times 1.7\, F(x)\, S_d \quad \text{...... (8.25)}$$

$$L = F(x) - 2 \times 1.7\ (1-F(x))\ S_d \qquad \text{...... (8.26)}$$

Where, S_d is the standard deviation of the binomial probability distribution $\left(Sd = \sqrt{F_x \frac{(1-F_x)}{N}} \right)$; $F(x)$ is the cumulative frequency and N is the number of data set.

Chapter - 9

Soil-Plant-Atmospheric Continuum

Before discussing about the Soil-Plant-Atmospheric Continuum (SPAC) let us understand about the potentials of soil water at which water is held. Soil water is classified into three types *viz.* gravitational water, capillary water and hygroscopic water. Gravitational water is that component of soil water which is drained out of the soil pores due to the influence of gravity. Capillary water is the soil-water component which is retained by the soil under pressure and may be available to the plants. This component of water is also termed as available water (AW). Available water is defined as the water amount held in soil system between the permanent wilting point and field capacity. Though field capacity is considered to be the upper limit of water available to the plant, the concept of field capacity in fact found to have lost favour from scientific community due to its ambiguity. The definition of field capacity indicates about the water retained by the soil when the drainage is complete. It was assumed that when the drainage of water markedly ceased, the soil water attains equilibrium and no more water is drained due to gravity. The water retained in the soil pores due to capillary force or metric potential is available to the plants till the metric potential reaches a value around 15 bars. However, this value may vary depending upon the types of plants. Filed crops generally have wilting point at 15 bars. Plant system is capable of withdrawing water from the soils through its roots to meet the demand of evapo-transpiration when water available in the soil is between these two ends.

Water along with dissolved minerals enters into the roots through a process called osmosis. What is osmosis? Osmosis is a process occurs when two solutions with different solute concentration is separated by a partition called semi-permeable membrane. A semi-permeable membrane is something like a curtain which allows only solvent to pass through but not the solutes. When two solutions with different concentrations are separated by the semi-permeable membrane, due to different concentration they tend to get equilibrium by sharing their solutes. As the semi-permeable membrane does not allow the solute to pass through, the solvent starts flowing across membrane to dilute the solution with higher concentration to make potentials equilibrium. This process is called osmosis.

Generally, the plant cell saps have higher concentration of salts. The soil water which has lower concentration is separated by a semi-permeable membrane in the roots. In order to equilibrate the concentration of salts, the water from the soil system enters into the roots. The balance of salts between the cell saps and soil water never happens as plant constantly losses water through transpiration at the leaves. The source of water for any plant system is at the root zones, whereas the sink is at the stomata of leaves through which plants give away water to atmosphere due to transpiration. From source to the sink there is a continuous movement of water due to the potential difference. Since the atmospheric potential is at about -1000 bar and root system is at 0.3 bar, naturally water will move from higher potential to lower potential and as a result water moves from the roots to the leaves through the innumerable xylems present in its body. Thus the process of movement of water from the soil to the atmosphere has continuity through the plant and hence this process is called soil-plant-atmospheric continuum or in short SPAC.

SPAC can be explained with the example of a wick in oil lamp where oil is kept in the container of the lamp and one end of the wick is dipped into the oil, the other end of the wick is burnt. The oil passes through the wick continuously from the container to the other end where it is burnt. This happens against the gravity but along the gradient of potential difference between the source and the sink. However, the similarity ends here. In the example, the process is passive unlike the SPAC. In case of plants, roots can expand or move deeper in search of water and at the same time the plants can have modification based on water availability. In case of water shortage plants can shed its leaves, close its stomata, change the size of leaves and change the orientation to sunlight or roll the leaves to reduce transpiration. The plants can also response to osmosis by changing its physiology. A schematic diagram of potential difference from roots to leaves is shown in Fig.1.

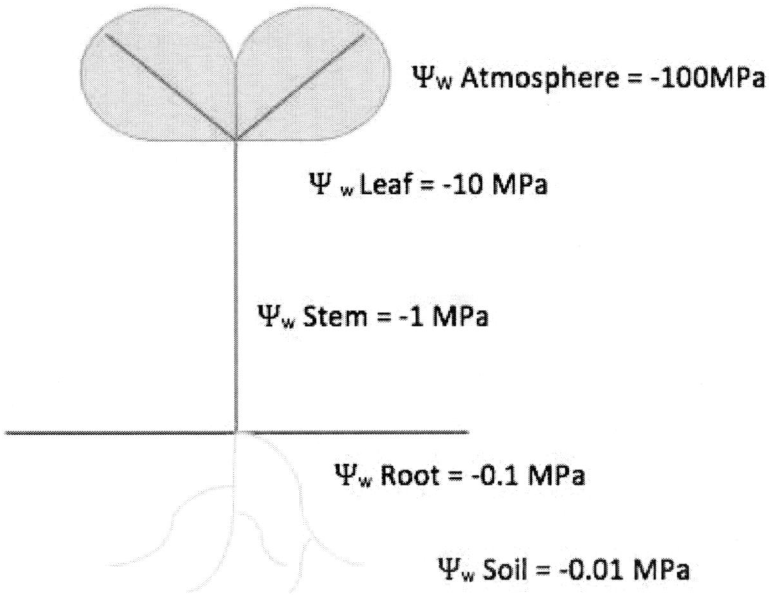

Fig. 9.1: Potential difference from roots to leaves in SPAC

Water is conducted through the plants either in liquid form or in gaseous form. The process is may be mass flow or diffusion. In these process, the water has to overcome hydraulic resistance in mass flow and diffusion resistance in case of diffusion process. From the Darcy's law, the mass flow in the SPAC can be written as:

$$q = -\left(\frac{1}{R}\right)\Delta\varphi \qquad \text{...... (9.1)}$$

Where q is the mass flow and R is hydraulic resistance and $\Delta\varphi$ is the potential difference. Negative sign is used because the movement is from higher potential to lower potential.

In case of diffusion process, the water movement can be expressed as:

$$J = -\left(\frac{1}{R*}\right)\Delta p \qquad \text{...... (9.2)}$$

Where R^* is diffusion resistance and reciprocal of Diffusion Coefficient D.

Both resistances, hydraulic and diffusive, increase in proportion to the transport distance, and decrease in proportion to the size of the transport parameters K(Ψ)(hydraulic conductivity) and D. For water flowing from soil to root, the drier the soil and smaller the K, the greater is the hydraulic resistance. But plants can modify the hydraulic resistance in the soil. By increasing root length density, plants reduce the transport distance from a position in the bulk soil to the root surface and thereby decrease the resistance. The variable resistances in the pathway of water from roots to leaves are given in Fig. 9.2.

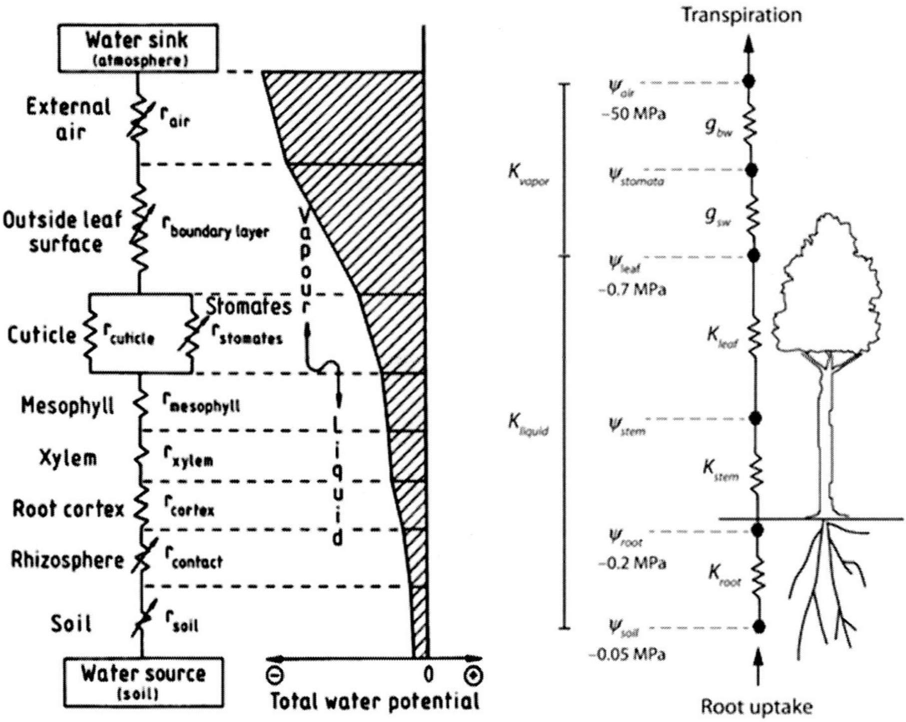

Fig. 9.2: Variable resistances form source to sink in SPAC (After Hillel, 1980)

In Fig. 9.2, the contact resistance depends on the soil water content, length and density of root hairs, swelling and shrinking of root diameter, all of them influence the total soil and root contact area at the transition of soil to root. The radial resistance of the pathway from the root surface through the cortex, including the endodermis, to the stele is considered to be much larger than the axial resistance in the root xylem. Also, the resistance in the xylem of the shoot is comparatively small. For vapour phase movement there is a considerable resistance at the cuticle and variable resistance in the stomata. The boundary layer resistance also changes based on the air movement around the leaves.

The reduction of water potential within the SPAC can be explained with four hypothetical situations. Firstly, the soil is moist and the potential at mesophyll tissue is much above the critical limit. Secondly, the transpiration rate is greater at identical soil water potential, so that the critical potential within the leaf is almost attained. Third situation is that soil moisture content is greatly reduced and the potential of some of the mesophyll tissues are at critical limit and the transpiration is also reduced. The fourth scenario is that potential falls below the critical limit and the plants wilt. All four hypothetical scenarios are shown in Fig. 9.3.

Under a given situation of evaporative demand of the atmosphere, the moisture potential within the SPAC is governed by the soil water potential. Because of contant uptake of water by the roots the matric potential and soil water content steadily decrease affecting the SPAC process. This is more pronounced when water supply is limited and water is extracted from only a limited soil volume. This becomes more conspicuous when plant is raised in a pot with limited soil volume. At mid-day, the high rate of transpiration often not balanced by the uptake of water by the root system.

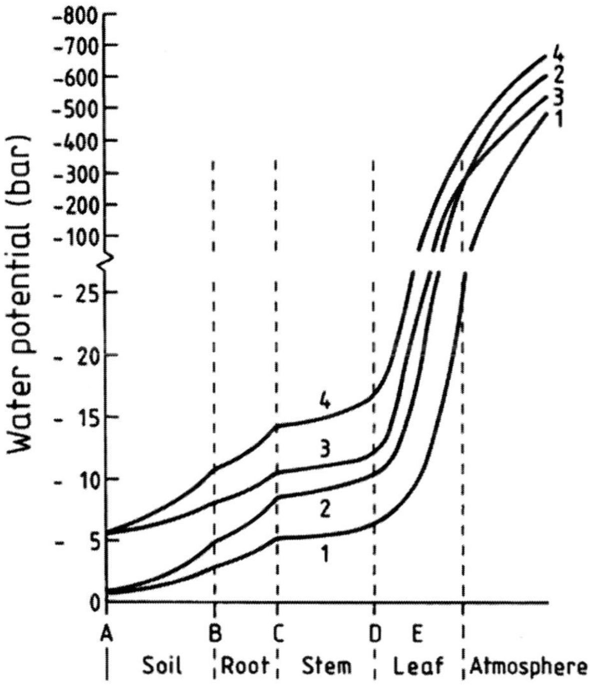

Fig. 9.3: Potential decline in SPAC at two soil moisture levels (After Hillel, 1980)

This causes water loss in the plant tissues. The loss induces the plant water potential to decline which the increase the hydraulic gradient in the system to trigger the water flux. The fall in water potential in the leaf is continued throughout the hydraulic pathway in the plant down to the root surface. The potential difference between root and soil will increase at noon, but not as much as between leaf and root surface in the plant. In the afternoon the transpiration gradually decreases and the loss of water reduced as compared to uptake as a result the tissues further get watered as overall water potential in the plant body increases. By the end of the night, a balance is usually achieved. At mid-day due to decline of water potential in the plant, the cell turgor is also reduced affecting the rate of cell enlargement. In restricted water supply such decline of growth hence becomes evident. In the afternoon, the cell enlargement and the growth increase again. The cell elongation continues through the night. Therefore, the growth at night is found to be always greater than that in day time.

Soil hydraulic conductivity is another property which influences the SPAC process. At lower hydraulic conductivity, often water flow is reduced and fails to supply sufficient water to the crop to meet the evaporative demand. Therefore, in the highly compacted soils or heavy clay soils, the stomata openings are restricted to reduce the transpiration by the plant. Similarly, in the sandy soil where the water holding capacity is too low, water cannot be supplied sufficiently to meet the demand of water transpiration. This not only due to low water availability, in such soil the root length is also limited resulting marked changes in plant physiology.

Chapter - 10

Controlled and Protected Cultivation

Food security is followed by demands for food quality. Further, the demand for higher quality food also necessitated the production of more food per unit area. This duel demands can be possible to meet only when the agriculture environment is under controlled management with optimal and precise inputs. Crop environment includes light, temperature, water, and minerals. Controlled environment is not only capable of providing these materials in adequate quantity but also can provide the most suitable environment where photosynthesis may occur at the maximum level. Photosynthesis is a process where plants produce carbohydrates in presence of CO_2, water, sunlight and minerals in optimal quantity. Plant cannot perform well if these are available at excess as well as in deficit proportions. Further, the variability of the environmental parameters also has effect on the health of the plants. Diseases, pest and insects all depend on the weather. If the crop is grown under controlled environment, the incidence of insects, diseases and pest is also reduced considerably or they are easy to control. With optimal supply of input without any pest and disease infestation, the produce is bound to be of good quality which can be achieved under controlled or protected conditions of agriculture. The controlled environment agriculture is also termed as protected agriculture. The example of protected agriculture is growing crops in the green houses.

10.1 What is greenhouse?

A green house is a framed structure made of bamboo, wood, pipes, truss, etc. covered with transparent or translucent materials and big enough to facilitate the movement of human being for undertaking cultural operations to cultivate crops. A green house is usually covered with materials like glass, polyethylene, acrylic, PVC, etc. films which allow the sunlight particularly

Photo synthetically Active Radian (PAR) to pass through and but do not allow the long wave radiation emitting from the earth surface to escape thereby increasing the temperature inside the house. Increased temperature and CO_2 in the green house may enhance the photosynthesis by up to 15 times higher than that in open environment. Originally, green houses were made of only glass to grow fruits like melons, grapes, strawberry etc. but gradually extended to the cultivation of vegetables. After World War (WW)-II, with the invention of polythene, the green house technology is established well in the commercial production of fruits and vegetables. Over the years, it evolved to a technology from mere plant protection to a technology of precision agriculture with fully automatic operations. Present day green houses have become rather factories of fruits and vegetable producing crops round the year with precise control of temperature, humidity, CO_2 and water. Almost every aspect of agricultural process is automated with minimum intervention of human being.

10.2 Green house effect

In Chapter 3, we have discussed how solar energy comes to the earth surface and benefits the all living beings. The solar radiation comes from the sun mostly in the visible range, about 95% of the total radiation occurs at the visible range of spectrum, due to the temperature of the sun, the same is not true for radiation from the earth surface. The average temperature of the earth can be considered as 300 K and according to the Wien's displacement law; earth can radiate energy at the wave length of around 10000 nm, which is in infrared radiation range of solar spectrum. The most of the thermal energy is radiated in the infrared radiation range of spectrum. Due to its longer wavelength with lower energy, the thermal radiation cannot escape from the earth's atmosphere due to its protective cover made up of gases like H_2O, CH_4, CO_2, N_2O, SO_2, etc. The protective cover reflects the thermal radiation back to the earth as thus the temperature of the atmosphere is maintained. This process is very important to maintain the atmospheric temperature. Had all the thermal radiation escaped the earth's atmosphere, it would have been very cold for survival of any organism. But excessive trapping of the heat energy is equally detrimental, which is being observed at present due to enhanced production of all the gases especially after industrialisation with high fossil fuel consumption.

The greenhouse is covered with transparent or translucent materials which are capable of transmitting the visible spectrum of sun's energy but trap the thermal energy within the house. The trapping of thermal energy within the house causes increase of temperature. Further, increased temperature under higher diffused sunlight, since the incidence light is much diffused while penetrating the greenhouse cover, leads to enhanced photosynthesis within the greenhouse.

10.3. Advantages of greenhouse

The greenhouse environment helps in increased photosynthesis. Higher photosynthesis results in higher crop productivity as well as enhanced biomass. Apart from increased productivity, a greenhouse brings up following advantages:

1. Cropping intensity can be increased to the maximum by taking 4-5 crops in a year
2. Maximizing the income per unit land area utilization in a year.
3. Quality of produce is found to be better due to minimum incidence of diseases and pest and insects.
4. Quality of produce is better due to effective input management because the cultivation is done under protective enclosure.
5. Off-season high value as well as exotic crops can be cultivated.
6. Requires less amount of chemicals
7. Nursery management is effectively done as the greenhouse ambience helps in improved germination and healthy seedlings.
8. Automation of agricultural operation such as irrigation, fertigation, humidification or dehumidification, temperature control, etc. is possible.
9. Multiple tiers of crops can be grown in per unit land area.
10. Greenhouse can be possible to use for drying of grains and other agricultural products.
11. Growing media can be used for different crops mitigating the possibility of development of soil borne pathogens. Soil borne pathogens can be easily developed due to humidity and high temperature prevailing in any greenhouse which necessitated soil treatment or solarisation. Sometimes soil borne pathogens compel to change the soils in the greenhouse after 3-4 years period of continuous cultivation.

10.4. Requirement of Green House Environment for Crop Growth

Crop productivity is undoubtedly influenced by genetic makeup and the available nutrients but at the same time the influence of micro-climate is very prominent. The micro-climates include light, air composition, temperature and the nature of rhizosphere. In open field, one can manipulate the nutrients, water, tillage, etc. but the other parameters of micro-climate are not possible to control. Green house environment gives the opportunity to manipulate these parameters including the air composition so that crop productivity is improved considerably.

10.4.1 Light

Light provides the energy to complete the bio-chemical reaction through which carbohydrate is produced from CO_2 and water in presence of chlorophyll. The photosynthesis reaction can be represented in the following way:

$$CO_2 + \text{Water} + \text{Light Energy} \xrightarrow[\textit{Plant Nutrients}]{\textit{Chlorophyll}} \textit{Carbohydrate} + \textit{Oxygen}$$

$$3CO_2 + 6H_2O \xrightarrow{\textit{Sunligh Photosyntnesis process}} C_6 + H_{12}O_6 + 3O_2$$

The energy carried by a photon, the unit of light particle, is called one quantum. Such 12 quanta are needed to produce one carbohydrate molecule. Hence, the amount of energy must be adequate to produce sufficient carbohydrates for crop growth and development. However, excessive light energy does not translate into higher production of photosynthate. Most of the crops become light saturated when the light intensity is 32.2 klux. Light saturation means that photosynthesis does not increase in the crops beyond the intensity. However, this limit greatly varies depending upon the crop and their growth stages. In green house environment, the light intensity can be ranging from 129.6 klux in a clear summer day to 3.2 klux in a cloudy winter day. Hence, one needs to understand the light requirement of the crops that grown under green house.

10.4.2 Role of light

Generally, plant captures a small spectrum of light in the range in the red region at 600-700 nm (nano meter). Light spectrum above it results weak and long plants and range below 400 nm suppresses the growth. The chlorophyll pigment present in the plant leaves captures light to be used for the process of photosynthesis. There are two types of chlorophyll that exist in the photo systems of green plants and they are chlorophyll "a" and "b". Chlorophyll "a" absorbs energy from wavelengths of blue-violet and orange-red light at 663 nm.

Chlorophyll "b" absorbs energy from wavelengths of red blue light at 645 nm. It is the accessory pigment that collects energy and passes it on to chlorophyll "a". It also regulates the size of antenna and is more absorbable than chlorophyll "a". Chlorophyll "b" complements chlorophyll "a". Its addition to chlorophyll "a" increases the absorption spectrum by increasing the range of wavelengths and broadening the spectrum of light that is absorbed. When there is little light available, plants produce more chlorophyll "b" than chlorophyll "a" to increase its photosynthetic ability.

10.4.3 Artificial source of light

In green house production the light can be controlled by providing artificial sources to overcome the limitations of cloudy weather. The use of artificial lights has been reported in 1870. Bailey (1981) in a study observed, early maturity of lettuce by nearly 2 weeks under cloudy conditions. Artificial lighting are required to supplement the natural light so as to extend the photo period. Generally light requirement varies from 5-50 W/m2 of the growing area.

10.4.4 Selection of light source

There are broadly four categories of lights used in green house they are incandescent, high intensity discharge (HID) fluorescent, high intensity fluorescent and light emitting diode (LED) lights.

10.4.4.1 Incandescent

Incandescent lamps are simplest types of lighting systems used in greenhouses. These lights are available in the range of 75-150 W and can be used for supplement light level within the system. The conversion ratio of light to energy is as low as 7% (lumen/watt). The other disadvantage with these light is it generates excessive heat to obtain desired light levels.

10.4.4.2 Fluorescent

Fluorescent lights provide a satisfactory alternate for lighting in green houses. These lights have higher light efficiency with low heat. Generally, a mix of cool and warm bulbs to provide full spectrum for different crop growth stages. The conversion rate of these lights are 40-60 lumens/watt. It converts 20% electrical energy into light energy. One of the advantage of fluorescent light is it provides uniform light. However, more light fixtures are required thereby increase the weight on the structures. These are available in the power rating of 125 watts.

10.4.4.3 High Intensity Discharge Lamps

High intensity discharge lamps: There are three types of high intensity discharge lamps they are high/low pressure sodium lamps, metal halides and high pressure mercury lamps. HID lamps convert electrical energy into photo synthetically active radiation very efficiently. They produce light in the blue and red spectrum that are beneficial to plant growth. These are compact and require less maintenance. Remote ballasting is possible, thereby, reducing the heat dissipation in cropping area.

A. High and low pressure sodium lamps

Pressure Sodium Lamps are mostly preferred by many growers for supplemental lights in greenhouse. The light produced is in the red/orange spectrum, which is especially beneficial in promoting flowering/budding in plants. These are available in 400 & 1000W, convert 25% of electrical energy into visible light. The Low pressure produces yellow light; they can convert 27% energy into visible radiation. The life of these lights can be as high as 20, 000 hours.

B. Metal halide

Metal halide lamps are better for use as the primary light source for plants in settings without much natural sunlight. The light produced is weighted in the blue spectrum, which is especially beneficial in promoting plant growth. These are available up to 2000W size and can convert energy up to 20% visible light. Limitation of these lights are costly, produce heat and low life. Combination of HID lamps with high pressure sodium is recommended. However, these lights require heavy ballast.

C. High pressure mercury lamps:

High pressure mercury lamps produces light due to electrical discharge between two electrode in a high pressure mercury vapour contained in a transparent bulb. These light produces light of high luminous intensity. These lights are available in the size of 400 and 1000 W.

High-pressure discharge lamps have been the preferred light source for crop production in controlled environment agriculture. The high PAR emission with relatively high percentage of blue radiation, long life span, and the electrical efficiency in the range of 25-40% make these lamps an option to replace daylight totally or partially supplementing it for year-round cultivation (Simpson, 2003).

A. **High and low pressure sodium lamps**

   Pressure Sodium Lamps are mostly preferred by many growers for supplemental lights in greenhouse. The light produced is in the red/orange spectrum, which is especially beneficial in promoting flowering/budding in plants. These are available in 400 & 1000w, convert 25% of electrical energy into visible light. The Low pressure produces yellow light; they can convert 27% energy into visible radiation. The life of these lights can be as high as 20, 000 hours.

B. **Metal halide**

   Metal halide lamps are better for use as the primary light source for plants in settings without much natural sunlight. The light produced is weighted in the blue spectrum, which is especially beneficial in promoting plant growth. These are available up to 2000W size and can convert energy up to 20% visible light. Limitation of these lights are costly, produce heat and low life.

Combination of HID lamps with high pressure sodium is recommended. However, these lights require heavy ballast.

C. **High pressure mercury lamps**

High pressure mercury lamps produces light due to electrical discharge between two electrode in a high pressure mercury vapour contained in a transparent bulb. These light produces light of high luminous intensity. These lights are available in the size of 400 and 1000W.

High-pressure discharge lamps have been the preferred light source for crop production in controlled environment agriculture. The high PAR emission with relatively high percentage of blue radiation, long life span, and the electrical efficiency in the range of 25–40% make these lamps an option to replace daylight totally or partially supplementing it for year-round cultivation (Simpson 2003).

10.4.4.4 Light Emitting Diode (LED)

The LED lighting is new application in greenhouse lighting. One of the major advantage is smaller size and easy to mount thereby reducing the load on the structure. Other advantage is it produces very less heat and have long life. These light produces blue and red spectrum that is good for plant growth. The energy consumption is around 80% less than HID lights and 30% less than fluorescent lamps. It provides immense prospects in green house lighting as claimed by its manufacturers.

The drawback of artificial lighting is rise in heat within the enclosed chamber, obstruction of natural light due to fixtures and increase of dead weight on the green house structure. Selection of suitable light is necessary to maintain the heat loss within the system. The amount of heat generated by lamp could be determined.

$$\text{Heat (J/s)} = \text{Watt of lighting including ballast} \times \text{eff. Factor}$$

[eff factor 0.75-0.8]

10.4.4.5 Air composition

As shown in equation discussed earlier in section 10.4.1, for triggering the process of photosynthesis plant requires CO_2 (Carbon di oxide). CO_2 uptake by green leaf in presence of sunlight or supplemental artificial illumination is a common measure of photosynthetic rate. The CO_2 in the presence of energy (sunlight) produces carbohydrate and oxygen. The CO_2 assimilation by plant can ultimately be translated into growth, yield and quality.

In the atmosphere the CO_2 concentration is only 0.035% (350ppm). In green house (closed one) the CO_2 concentration is variable with higher concentration during night that reaches up to 1000 ppm and during day time in presence of light it starts to fall gradually. The concentration may fall upto 300 ppm or below by afternoon and may become deficient. Under such situation the photosynthesis process may get affected if not augmented.

The concentration of gaseous component is generally expressed in terms of mass-volume concentration Pc, usually as micrograms of CO_2 per cubic meters of air plus CO_2 ($\mu g/m^3$)

10.5 Factors affecting CO_2 uptake

The uptake of CO_2 is affected by several factors. Some of these factors can be listed as below:

a. Plant species and variety

b. Radiation intensity

c. Wind velocity

d. Water stress

e. CO_2 con. in air

f. Resistance to CO_2 diffusion through the stomatas.

g. Previous history of plant and

h. Leaf area

i. Temperature (chemical reaction commonly be doubled for energy 10 °C rise in temperature over limited range)

At normal CO_2 levels, maximum photosynthesis of individual leaves of many species takes place at light intensities around 21530 lux. As CO_2 increase maximum intensity for photosynthesis also increases.

10.5.1 CO_2 enrichment

As the photosynthesis process begins with day light, the concentration of CO_2 starts falling. Under such circumstances enrichment of CO_2 have to be done to maintain the rate of photosynthesis. Enrichment process can continue till one hour before sunset. Enrichment is generally not practices in ventilated green house. Enrichment process may lead to rise in temperature and thus should be avoided in summer.

10.5.2 Methods for enrichment

The enrichment process can be achieved by either combustion of hydrocarbon, liquid CO_2 or solid CO_2. While selecting the fuel the quality in terms of quality viz., sulphur content ≤ 65mg per 30 m^3 of gas should be kept in mind. Since suitable fuel has to be used for enrichment it also involves cost. The cost can be calculated from the annual duration of combustion.

The quantity of CO_2 enrichment need to be estimated based on crop requirement and loss through infiltration. As a thumb rule the requirement of CO_2 varies from 0.6-1.2l/hr/m^2 of floor area.

10.6 Temperature

For the best performance of the crops, it requires to be grown within its cardinal temperature. Depending upon the crops the cardinal temperature can greatly vary. When temperature becomes lower than the minimum temperature, the life process comes to standstill and at the same time when temperature becomes more than the cardinal temperature, the enzyme required for bio-chemical reactions become inactive. Below the ceiling temperature of any crop, the rate of bio-chemical reaction often becomes doubles or triple for per 10 °C increase of temperature within a limited range. Hence, more the temperature higher is the photosynthesis till the maximum cardinal temperature reached. In green house environment however, temperature ranging from 15 °C to 28 °C is considered optimum depending upon the crops. As a thumb rule, green house crops are grown at a temperature 3-6 °C higher than the ambience at night and about 8 °C higher during clear day.Increase in temperature enhances transpiration process. If the soil temperature drops it affects the water and nutrient uptake irrespective of air temperature. It may impact the plant elongation too.

In green house cultivation, temperature at which plant can grow depends on various factors.

i. Species, cultivar and clone.

ii. Available radiant energy (light) which is influenced by species, planting arrangement, density, green house structure and its cover and climatic condition.

iii. Water availability

iv. CO_2 conc.

v. Stage of growth, age, and the particular plant structure

vi. Nutrition of crop, fertilizer application and water quality

10.7 Humidity

Relative humidity affects leaf area development and stomatal conductance there by interfering with photosynthesis and dry matter production.As the greenhouse environment is a closed environment, the Relative Humidity within the green house is expected to be higher than that in the ambience as the moisture will be added by the evapo-transpiration process. A green house may have acceptable range of RH between 50 to 80%. However, for plant propagation, the required RH may be about 90%. In order to maintain the humidity, a greenhouse may require humidification or dehumidification processes.

10.8 Types of Greenhouse

Green houses can be classified based on four criteria *viz.* type of climate control whether partial or complete, type of cladding materials used, type of shape of the structure and the purpose of the structure. A greenhouse can be made as a complete climate controlled or partially controlled depending on the purposes or investment by the farmers.

10.8.1 Classification Based on Shape

Though green houses are just framed structures covered with cladding materials, certain requirements make the structures varying in sizes and shapes. Purposes of the structures, light requirement, necessity to cool or heat, topography, slope and slope aspect, materials of construction, crops to be grown, ease of operations, etc. are few points which decide the shape of a greenhouse. Green houses may be classified based on its uses or based on methods used for maintaining the climate within it. Fig. 10.1 and Fig. 10.2 shows different classifications of greenhouse based on its uses and its maintenance. Most commonly adopted shape of the houses are:

a. Lean-to Type
b. Gable even type
c. Uneven span type
d. Quonset type
e. Ridge and Furrow type
f. Saw tooth type

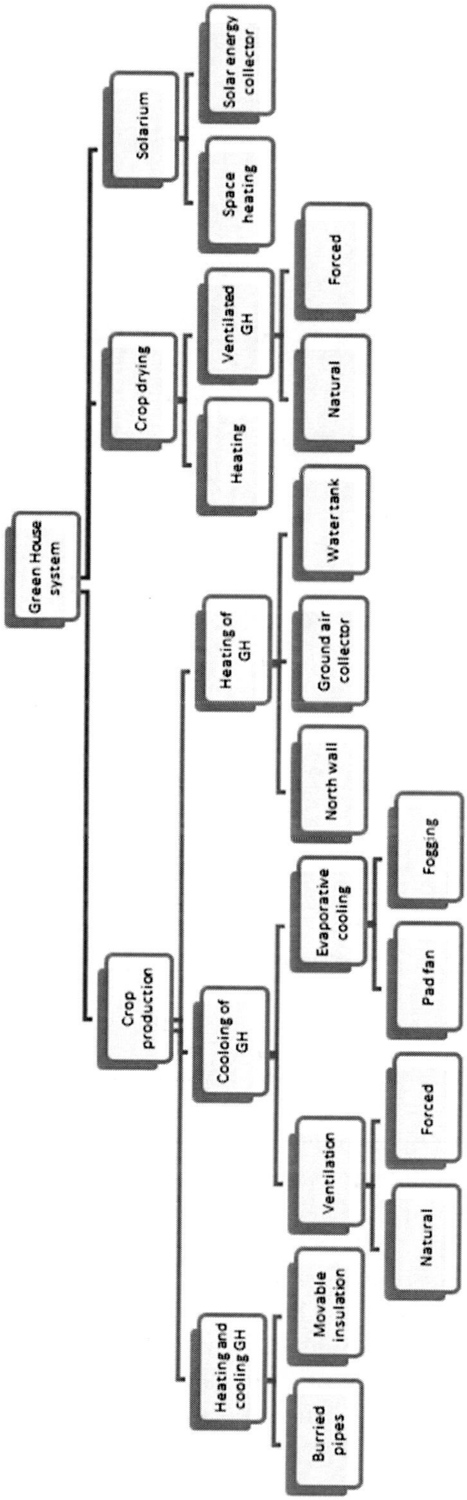
Green House system
Solarium
Solar energy collector
Space heating
Crop drying
Ventilated GH
Forced
Natural
Heating
Crop production
Heating of GH
Water tank
Ground air collector
North wall
Cooloing of GH
Evaporative cooling
Fogging
Pad fan
Ventilation
Forced
Natural
Heating and cooling GH
Movable insulation
Burried pipes

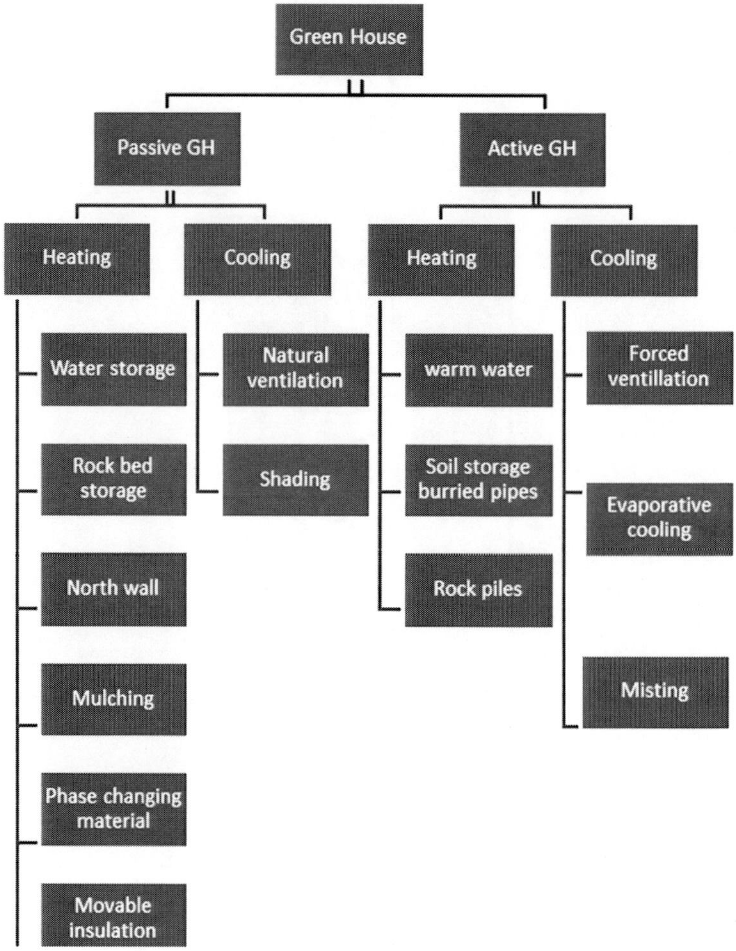

Fig. 10.2: Classification of Greenhouse depending on its maintenance

10.8.1.1 Lean-to Type

These types of green houses are single span green house and constructed against existing structure. The span is directed to the sunlight (southern direction in northern hemisphere and vice-versa) and makes the best use of sunlight. As the direction opposite to the sunlight is the existing structure, the northern side of the house (in northern hemisphere) is usually insulated and as result energy loss in these types of house is very low. Such greenhouses can also be used for warming the residential houses in cold regions. If a lean-to type green house is constructed against the wall of any existing houses, the heat inside the house also keep the adjacent rooms of the house warm. The cost of construction of such greenhouse can be reduced considerably. Such structures are however preferred when the size of the green houses are small (Fig. 10.3).

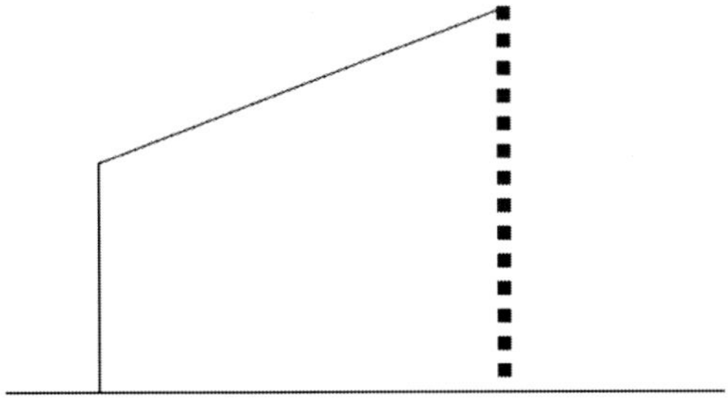

Fig. 10.3: Lean-to type structure

10.8.1.2 Gable Even Type

This shape is also adopted for small green house. Generally it is constructed on a level ground and the both the spans of the house are equal in size and slope (Fig. 10.4). The length of the span form single span of the house is about 5 to 9 m and the length of house is usually kept within 30 m. The heights of such houses are 2.5 m to 4.3 m. This type of green houses are also called as A-frame

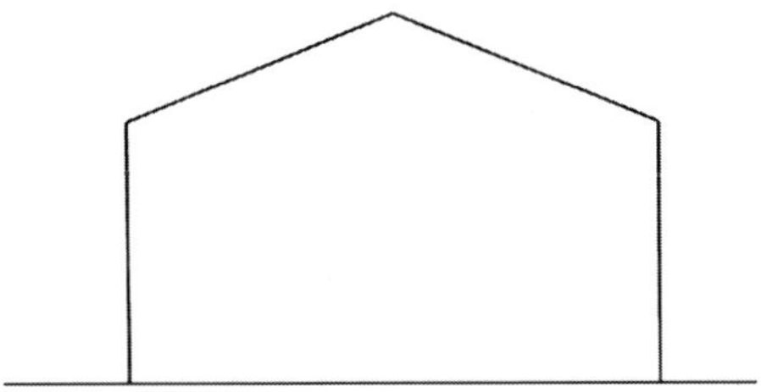

Fig. 10.4: Gable Even type of Structure

10.8.1.3 Uneven Span Type

Uneven span type greenhouse is suitable for hilly region, where the slope of the topography is utilized in such a way that the span exposed to the sunlight is made longer. The total span area exposed to the sun facing side is maximised to get higher sunlight. This type of structure does not require levelling of the land and the cultivated land is usually terraced (Fig. 10.5).

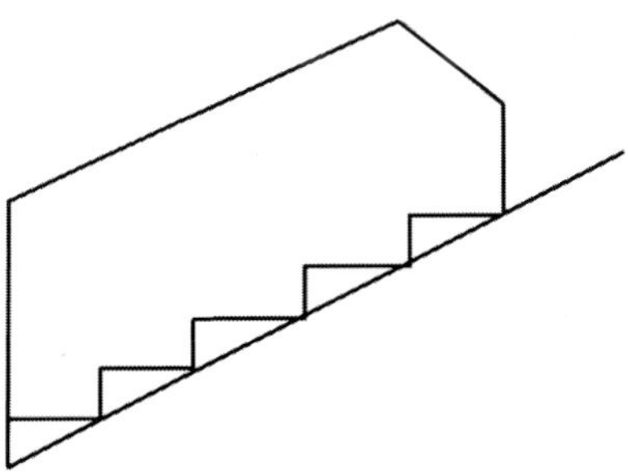

Fig. 10.5: Uneven Span type of structure

10.8.1.4 Quonset type

In Quonset type of green houses, the trusses are usually made of metal arches and supported by purling along the house (Fig. 10.6). When isolated houses are needed, this type of house is better due to low cost involved. Low cost greenhouses made of polyethylene are constructed in this shape. Quonset type houses can be either free standing style of house or interlocking type with multiple spans having the interlocking portion as gutter.

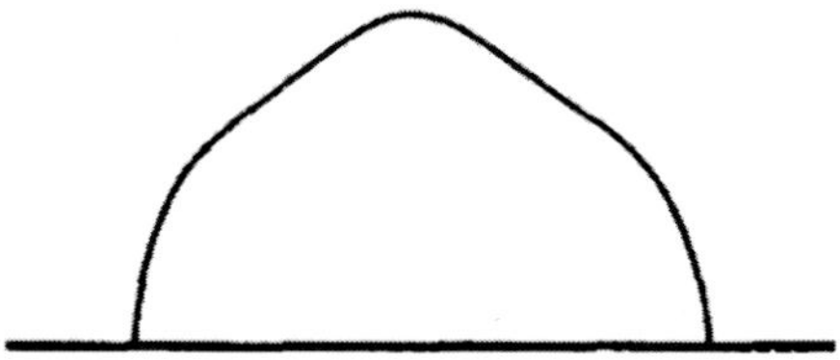

Fig. 10.6: Quonset type of structure

10.8.1.5 Ridge and Furrow Type

When standalone gable even houses are interlocked to make multi-span house, a ridge and furrow type of structure is formed. In ridge and furrow type of green houses, the edge of each house connected to each other to crate the gutter or furrow. Since the side walls of the house can be removed, a large cultivated area can be covered with such structure. Due to consolidation of cultivated area under a single roof can enable one to reduce input cost for cultivation practices. Automation of such green house is easier due to single management domain. While designing the house, it is necessary to take care of water load in the gutter (furrow) in high rainfall region and snow load if it is constructed in cold region (Fig. 10.7).

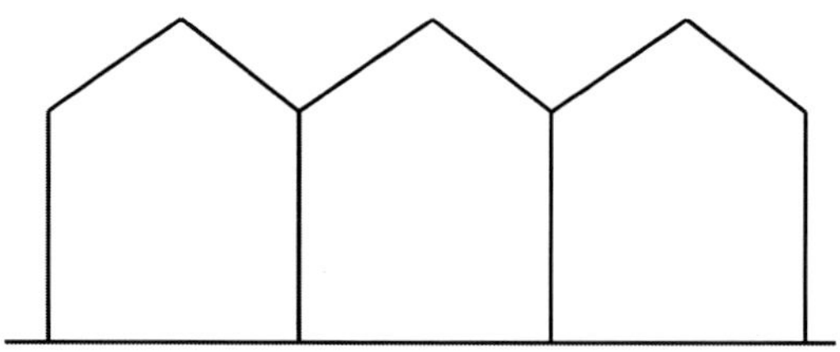

Fig. 10.7: Ridge and Furrow Type of structure

10.8.1.6 Saw tooth Type

These are also multi-span green house similar to the ridge and furrow but only difference is that these are constructed in such a way that it has ventilation at the top of the house in the direction opposite to the sunlight. The span facing to the sunlight is made larger similar to the slopping sides of the saw tooth. A very good air circulation can be achieved in such type of structures and natural draft of cooling the house can be effectively utilized (Fig. 10.8).

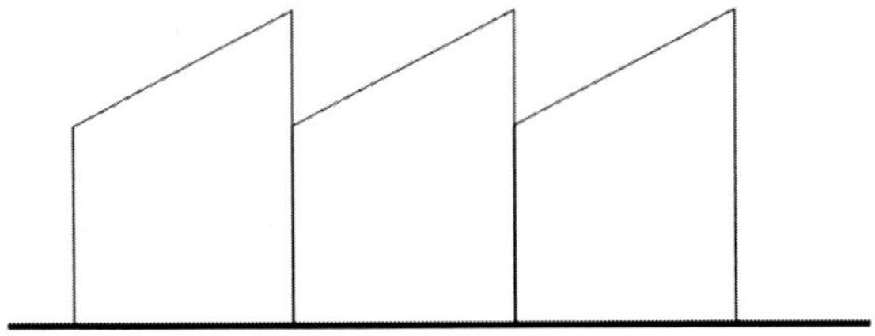

Fig. 10.8: Saw tooth type structure

10.8.2 Classification Based on Type of Climate Control

The optimum temperature range in a green house should be kept at 15-28 °C, which is also cardinal temperature of most of the crops. If the ambient temperature goes below the temperature, it becomes necessary to raise the temperature so that crops do not suffer. Similarly, at temperature higher than the 28 °C, a greenhouse needs to be cooled. Depending upon the requirement of heating or cooling at different regions, a house can be classified for

a. Active cooling/ heating and

b. Passive heating/ cooling

When temperature is anticipated to be higher than the cardinal temperature, the houses are necessary to be designed to accommodate cooling mechanism. Cooled air circulation into the house by replacing the hot air can be one method of cooling of a green house. This usually require higher and regulated ventilation. Apart from natural draft cooling through lair circulation, one can go for evaporative cooling mechanism in a green house. This is similar to the application of desert cooler as used in the summer days for dry and hot regions of India. If the farmers can invest high amount for active cooling house, air-conditioning system may replace the evaporative cooling mechanism.

In winter period or in the regions of high altitude and temperate climate, the ambient temperature goes below zero degrees Celsius. In such situation the temperature within the green house also may not be suitable for crop cultivation. Active heating type of green houses are made for good crop response. Different types of heating mechanisms are provided which will be discussed later.

10.9 Types of Cooling system

During summer there is always a increase in temperature within the green house. It is necessary to bring down the temperature for providing required temperature for the crop production. Different systems are applied to bring down the temperature within the green house. Different cooling concepts will be discussed in the following sections:

10.9.1 Passive Cooling

10.9.1.1 Natural draft cooling

This is a passive cooling system. It is suitable for summer cooling process. In this method no external force is applied to enforce any cold air but the natural draft is used to pull cold air and replace the warm air thereby reduces the temperature within the house. This cooling mechanism is feasible only when the ambient temperature doesn't go beyond 30 °C since the natural draft cooling system can make the green house temperature equilibrium to the ambience. It is a common knowledge that warm air is lighter than the cold air as a result given a passage, cold air can easily replace warm air from any enclosure. A natural draft cooling system can be made either by providing ventilation at the top of the house or by giving provision to open the side walls of the house to enable cross flow to remove warm air. In absence of any external force such as exhaust fans, effective natural ventilation is in fact a challenge for the designers (Fig. 10.9).

In natural draft cooling green house, ventilators are provided at the roofs and side walls. There must have provision to close and open the ventilators as and when required. When the internal temperature of the green house goes past 30 °C, the ventilators need to be opened. The warm air, which is light occupy the upper part of the house. As soon as the air outside can enter through the ventilators at the side walls, the warm air at the upper part of the house goes out through the roof-ventilator and the house get equilibrate with the ambient temperature. Total ventilation area should be about 10% of the roof area. During winter cooling phase, the south roof ventilator was opened in stages to meet cooling needs. When greater cooling was required, the north ventilator was opened in addition to the south ventilator. In summer cooling phase, the south ventilator was opened first, followed by the north ventilator.

Another mechanism is used for natural ventilation. This is more common in case of low cost polyethylene cladding green house where the side walls are rolled up to equilibrate the temperature of the house with the ambience. Roller mechanisms are usually provided while construction. Whenever it is necessary to reduce the temperature, the side walls are rolled up to allow the warm air to be pushed by the incoming cold air to bring down the temperature.

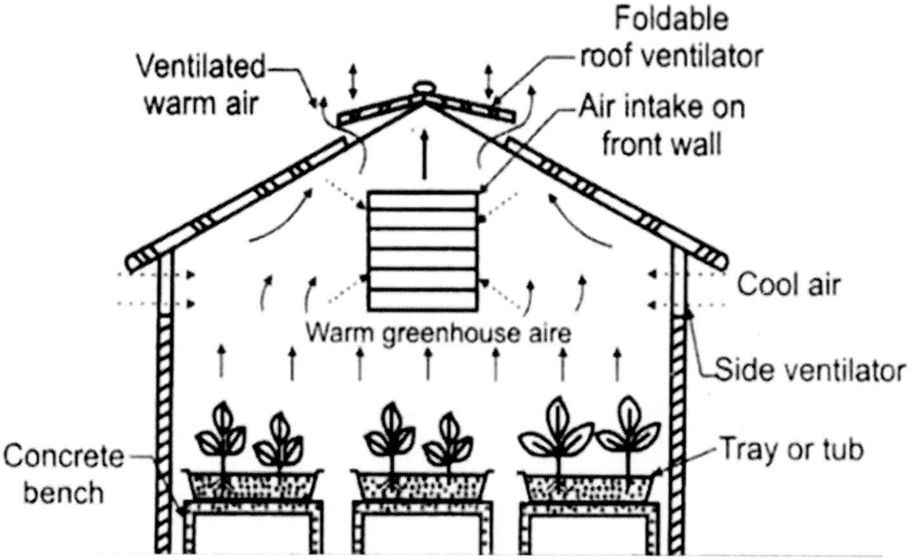

Fig. 10.9: Natural draft cooling with ventilation

10.9.1.2 Forced convections:

During summers when the rate of heating within greenhouse is higher than its removal, forced convection system may be adopted for heat removal. An exhaust fan can be installed to take out the warm air within the system. The rate of heat removal depends on the capacity of the fan and its rpm. Such system may be useful during the month of April and May.

10.9.1.3 Nocturnal radiation cooling:

It is a type of passive cooling. In this system radiative cooling is achieved during night time through roof materials that emits infrared (long wave) radiations. This results into heat loss and drop in temperature. This system is effective in hot dry climate and requires no water. In humid region it is not effective due to condensation from atmosphere takes place due to drop of temperature below due point. As a result, loss of heat is not realised. Similar effect is experienced on cloudy night also.

10.9.1.4 Cooling by night convection:

In region where night temperature in summer is lower than desirable temperature for plant growth (below 20 °C), it is possible to store the coolness of the night air. It can be achieved by putting storage material such as gravel or rock. The stored coolness can be used to cool the green house during day time.

10.9.2 Active Cooling System

10.9.2.1 Roof evaporative cooling:

Roof evaporative cooling is based on the principle of conversion of sensible heat to latent heat. In this process large amount of heat is removed. It can be achieved by placing wet cloths like gunny bags on the roof of green house. The material selected for covering should be thin as far as possible. When the solar radiation incidence on the wet surface, evaporation starts. The wind above the roof removes the humid air, but it does not heat the roof surface. Simple precaution such as leak proof roofs should be ensured. Such systems are effective during the month of May and June or during peak summer.

10.9.2.2 Pad-fan cooling system:

Evaporative cooling of green house is the most common method wherever natural cooling (passive or active) is not effective. This is similar to the desert cooler used in summer season in the northern region of India. In evaporative cooling, humidity of the hot air inside the green house is increased by inserting humid air where evaporation is induced through air circulation and thereby the temperature of the house is reduced. In fan and Pad method of cooling, one wall of the house is provided with pad made of wood fibre or similar cellulose materials which is continuously made wet by pumping water onto it. On the opposite wall exhaust fans are fitted. When exhaust fans are operated, air will pass through the wet pad which will be enriched with moisture while passing through the pad. The moist air then passes across the house and goes out through the exhaust and brings down the temperature of the house considerably.

This method is quite effective when the distance between the fan and pad is kept between 30-61 m. Normally, exhaust fans are also necessary to be fitted at an interval of 6 m to provide uniform air flow across the house. The thickness of the cellulose pad should be about 2.5 to 4 cm. Bigger thickness, up to 30 cm can also be used depending upon the materials. Normally an air flow rate of 45 cum/m^2 is necessary for effective cooling of green house.

The National Greenhouse Manufacturers Association (NGMA) indicates in its 2004 standards and guidelines for ventilating and cooling greenhouses that a rate of removal of 2.5 cum/m^2 of greenhouse floor is sufficient. This applies to a greenhouse under 305 m in elevation, with an interior light intensity not exceeding 53.8 klux and an air temperature rise of 4 °C from pad to fans. The rate of air removal from the greenhouse must increase as the elevation of the greenhouse site increases. The density of air decreases and becomes lighter with increase in elevation. The ability of air to remove solar heat from the greenhouse depends upon its weight and not its volume. Thus, a larger volume of air must be drawn through the greenhouse at high elevations than that is drawn through at low elevations in order to have an equivalent cooling effect. Similarly, the rate of air removal is also dependent on the light intensity. With the increase of light intensity, the thermal energy accumulation inside the green house increases which necessitates greater air removal. The excessive light intensity is sometimes controlled by providing shade over the house through removable shade net cover. Maintaining uniform temperature throughout the house is necessary. If the temperature difference between the fan and pad goes beyond 4 °C, more air flow is needed to flatten the gradient. Fig. 10.10 show a typical pad and fan cooling system.

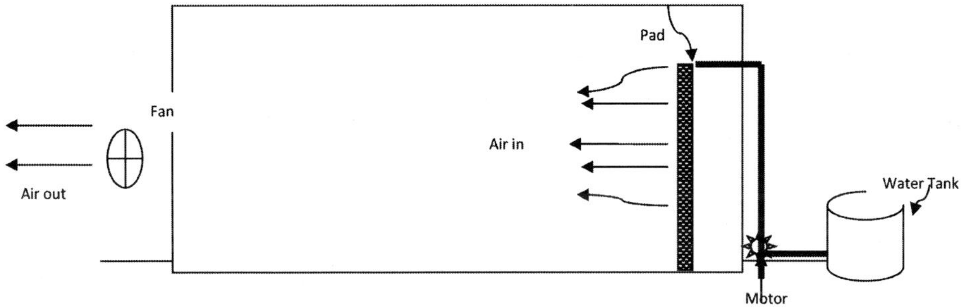

Fig. 10.10: Fan and Pad arrangement of Greenhouse Cooling

10.9.2.3 Foggers

Evaporative cooling is also possible by simply enriching the humidity of the air and allows to have forced circulation of the humid air. Mist blower or fogger in the green houses are generally used for that purpose.A high-pressure pumping apparatus generates fog containing water droplets with a mean size of less than 10 microns using suitable nozzles. These droplets are sufficiently small to stay suspended in air while they are evaporating and utilize the heat of greenhouse air. Fog is dispersed throughout the greenhouse, cooling the air everywhere. As this system does not wet the foliage there is less scope for disease and pest attack. People and plants stay dry throughout the process. This system is equally useful for seed germination and cutting propagation, since it eliminates the need for a mist system.

10.9.2.4 Movable screens (shading):

Screens can also be used to control the direct radiations entering the green house. Generally, these are put from north side to south side to cut off the radiations.

10.10 Green House Heating

The temperate climate in winter often needs heating system in green house to maintain desirable temperatures for growing crops. Even though greenhouse temperature is always higher than that in the ambience, heat loss is another common problem. To maintain desirable temperature, heat must be supplied as it is lost. Green houses are designed in such a way and with such cladding materials, that heat loss can be minimised. Heating of green house is a costly affair if it is not done with additional solar energy through some design modification. Burning fuels or using electricity to heat greenhouses often becomes prohibited for the farmers due to cost involved. In conventional greenhouses without any additional system of heating, some heat absorbing materials can be kept inside the house. Stones, sands and water are some common materials people keep to conserve some heat energy to keep greenhouse warm in winter places.

Heat loss from a greenhouse occurs through the three general process, *viz.* conduction, convection and radiation. Conduction process of heat transfer requires conductors or a medium through which heat can be transferred. The cladding material such a polyethylene, aluminium or steel frames, partition walls, etc. can be good means of conduction. In fact, there is limited possibility that heat loss can be minimised in normal conditions without blocking the radiation. However, providing double layer, instead of single layer cladding material, can minimise loss of heat by about 40%. For example, an A-frame green house with single layer of polyethylene may lose heat at the rate of 6.8 $W/m^2/h$ if the outside temperature is 1^0C lower than the inside temperature. The same will be about 3.97 $W/m^2/h$, if additional layer of polyethylene is provided.

Air infiltration can be another mode of heat transfer. This is also called convection. In this process the hot air itself will be exchanged with the environment for getting equilibrate. Apart from infiltration through leakages, air can be exchanged also while opening or closing doors of the house. About 10% of heat can be lost through convection in a structurally tight green house. In radiation process, heat is lost through emission of energy from the hot outer surface of the house. The radiation process follows the same principle as prescribed by Stefan- Boltzmann law.

10.10.1 Types of Heating System

10.10.1.1 Passive Heating System

Direct Gain:

In direct gain the sunlight is allowed to enter the greenhouse through its walls and roofs. The selection of covering materials used plays a important role in direct gain. To maintain the temperature at reasonable limits sometimes exhaust and vents are used particularly during summers. During winters shading may be provided to prevent loss of heat.

Indirect Gain

North Wall: The north wall is a thermal storage system placed on the north. This concept helps in preventing fluctuation in temperature within the system. These thermal storages may include water wall or trans wall. The water wall may be created by placing a water tank. The characteristics of water results in slow rise and fall of heat. Trans wall is a partially transparent storage wall placed adjacent to covering surface. The part of solar energy is absorbed and remaining is transmitted to greenhouse room air.

Isolated Gain: The working principle of isolated gain system consists of placing a collector of solar energy outside the greenhouse system. The energy collector consists of a pit filled up with materials such as sand, stones and concrete that absorbs the solar energy. The solar energy heats the materials used for collecting the energy. The heat is then conveyed inside the greenhouse with the help of tubes. The ground air collector is made of PVC tube and sand bed at 0.2 m depth in 6m x 2 m area.

10.10.1.2 Conventional Methods

The conventional methods are adopted in harsh climate. Conventional fuels viz., cow dung cake, coal cake, wood biogases etc are used. During burning CO_2 is produced.

a. Isothermal mass: In this case a cylindrical drum made up of metallic sheet with blackened surface, is filled with water and is placed inside green house. Solar radiation falling on the surface is absorbed by blackened surface and in turns heats the water.

b. Heat Pump: Using two heat exchanger for cold and warm water. For 24 hours thermal heating system from boiler is provided. An exhaust fan is also provided for ventilation to control humidity and CO_2.

10.10.1.3 Active Heating System

Generally three popular types of heating systems are adopted in green houses,*viz.* unit heating, central heating and radiation heating systems.

Unit heating system

This is the most common heating system used in green house as it is less expensive. In unit heating system warm air is blown from the fire place where wood or coal can be burnt. This system has three components- a fire place, heat exchanging tubes and blower fans. The fuel is burnt at the fireplace and the hot air collected from the fireplace is passed through exchanger pipes which are made of thin walled conducting materials. The metal exchanger pipes thus get heated up and the cool air passing in contact of the pipes becomes hot which raises the temperature of the green house air. Fan fitted behind the unit heater basically force the cold air to pass over the exterior of the metal exchanger pipes.

Central Heating System

This system is the most efficient one. In central heating system a boiler is kept at the centre of the green house along with a heat radiating mechanism. Water is boiled in the boiler and the steam can radiate most of the heat energy to keep the house warm. This system is generally very effectively used in large green house. In a central location, one or more boiler can be put considering the size of the house and amount of heat needed. Hot water is also passed through pipes fitted on the walls of the green house. The pipes are radiating type which can distribute the heat energy throughout the house uniformly. Pipes can be put in the soil or on the floor as required to raise the soil temperature also.

Radiation type

In this type of heating system gas is burnt inside the pipes suspended overhead in the greenhouse. Warm pipes due to burning of gases inside radiate heat energy, Low intensity infrared radiant heaters can also be used for warming green house. Infrared heaters can save energy by about 30% as compared to the conventional heaters.

Solar energy collectors are often fitted with green house to raise green house temperature. However, use of solar collector is a costly business.

Energy loss from green house can be minimised by operators if the following practice is adopted:

a. Tighten up the house by closing all possible openings
b. Don't frequently close and open the doors of the house to minimise the loss through convection.
c. Regular maintenance of heating systems particularly the hot air and water pipes is essential to avoid any leakage.
d. Reflector materials behind the exchange pipe to radiate maximise energy into the house is an important step.
e. Boiler efficiency generally get reduced due to scaling up of inside wall, which should be avoided to maintain the boiler at maximum efficiency.

f. Thermostat system should be regularly maintained.

g. Cladding should be of double layers, if polyethylene is used.

h. Removable thermal screen can be a good option to maintain the temperature of the green house both day and night.

i. Night screen over the crops or house can be put to minimise radiation of heat at night from the green house to the atmosphere.

Chapter - 11

Design and Construction of Green House

In India the construction of green house is done as per Indian standards IS 14462:1997 issued by Bureau of India Standards (BIS). The different aspects to be considered while construction of greenhouse are:

(i) Site selection

(ii) Layout

(iii) Orientation

(iv) Different loads on the greenhouse

(v) Methods of construction

(vi) Electricity requirement

(vii) Water availability

11.1 Site selection

The site selected should be such that there is adequate communication network for bringing in the inputs and transporting out the produce of the green house system. Efforts should be made such that it is away from polluting units and vehicular pollution that may affect the plant. It should be located away from trees and high rise buildings so that it gets adequate light. The land should be well drained by providing adequate slope for conveying out water if any. The green house should have south facing slope (less than 3%) for better illumination. Suitable wind breaks should also be provided for protection from damage and ensuring convective loss. In snowy areas the trees should be located at least 30 m away from the structure to protect it from shocking wind effect.

11.2 Layout

The greenhouse layout is another important aspect in its construction. A greenhouse consists of number of facilities such as space for work, storage of inputs, and area for sale apart from production space A typical layout of green house is shown in Fig. 11.1. The layout of all the facilities should be planned with following factors in consideration:

(a) The other spaces required for greenhouse production systems should be located on the north of production house. This layout ensures that the production house is free from shadows from other facilities.

(b) The storage facilities, receiving areas and sales office should be provided with separate entrance.

(c) Customers/visitors should not be allowed to pass through the production areas. They should be allowed with convenient parking space.

(d) Adequate wind breaks should be provided to protect from strong winds. Generally, they should be located at around 30 m distance from the greenhouse on north and west of the structure.

While constructing, the type of greenhouse need to be considered keeping merits and demerits in view. The green house may be single span or multispan depending on the economics in consideration.

A single span house may be kept separate or end walls attached to head house. The layout should have provision for future expansion. Separate greenhouse may have different micro climate. In single span house energy requirement for maintaining the climate within it is generally higher per unit floor space.

In multispan greenhouse centralized system can be used for maintaining the micro climate. Such systems are less costly as compared to single span houses. A minimum of 1000 m^2 area is considered adequate for regular maintenance within the house. Expansion of such houses may not be convenient as compared to single span houses.

While planning other aspects such as land availability, availability of water, electricity and instrumentation should be considered for future expansion.

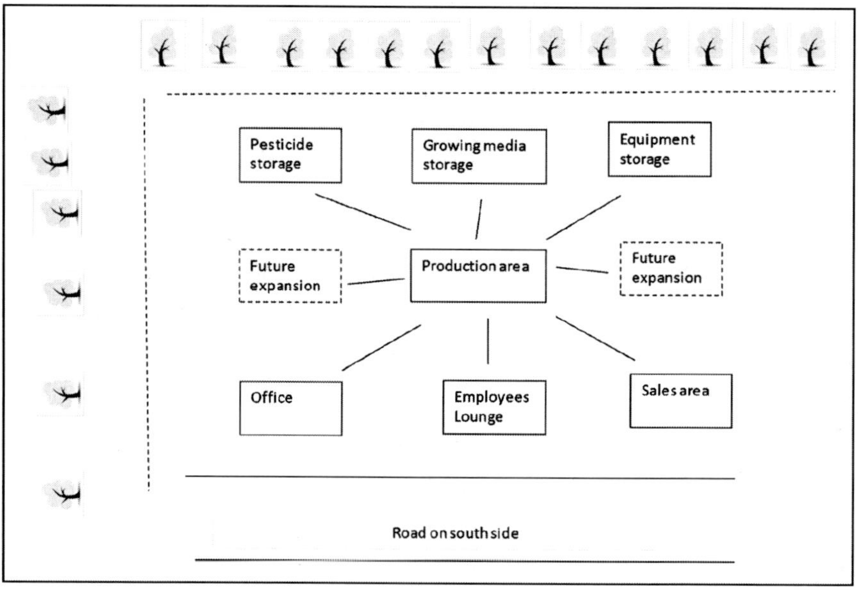

Fig. 11.1: Typical layout of greenhouse system

11.3 Orientation

Orientation is another aspect of greenhouse construction. Two criteria are considered for orientation of any green house

(a) There should be adequate and uniform light levels inside the greenhouse required for crop growth, and

(b) The direction of local wind if any should also be kept in mind so that it does not adversely affect the structure.

Depending on the type of green house (free standing or multi-span) the orientation should be decided. In free standing greenhouse the east west direction provides more light in winter thereby reducing the cost of heating. In multi-span green house with gutters North-South direction ensures better availability of light due to the drifting shadow.

11.4 Different loads on green house

Different loads (weights) acting on a greenhouse can be categorized as Dead load, live load and wind load.

11.4.1 Dead load

All the weights that are imposed on the green house such as Weight of permanent component like heaters, water pipe lines and all fixed service equipment. Apart from these loads long term crops such as cucumbers and tomatoes also fall under the dead load categories. For estimation of dead load, the actual mass of building materials and systems are taken into account.

11.4.2 Live load

Weights imposed on the structure due to its use and are temporary in nature are called live loads. These include weight of repair crews and hanging plants.

11.4.3 Snow Load

In snow covered areas the weight due to snow are also considered during design. In snow areas a minimum of 3 m distance between green houses to prevent from failure of walls due to snow sliding from the roofs. The snow load is estimated as prescribed in IS 87.5 (Part 4). A snow depth of 7.5 cm (wet) or 30 cm (dry) is equivalent to 2.5 cm depth of water.

11.4.4 Wind Load

The greenhouse should be designed to resist the wind load. The wind load of the greenhouse structures shall be estimated as prescribed in IS 875 (Part 3).

The effective wind velocity pressure can be calculated from the map by converting the wind speed using the formula (ANSI A 58.1-1972 6.3.4)

$$q = 2.37 \times 10^{-5} V \qquad \text{...... (11.1)}$$

Where q is the basic wind pressure in Pa; V is the wind speed in m/s; other factors such as gusts and height facility can be included in final calculation. The final wind load factor (w)

$$W = q \times q_p \qquad \text{...... (11.2)}$$

Where q_p is the external pressures acting at local positions.

As per the IS 14462: 1997 different loads on a greenhouse are shown in Table 11.1.

Table 11.1: Minimum design loads for greenhouse mainframes

Load description	Minimum value (N/m²)
(a) Dead - Pipe frame, polyether cover	100
- Truss frame, lapped glass	250
- Supported crops- tomatoes, cucumbers etc.	200
(b) Live - Workers, repair materials	250
(c) Snow - 10C minimum green house temperature	750
(d) Wind load acts perpendicular to surface	500

11.5 Construction of Greenhouse

The points to be considered doing construction of greenhouse includes foundations, flooring, frames, structural components and covering materials

11.5.1 Foundations

As per the IS specification pier foundation is adequate for a greenhouse construction. If the hoops are closely spaced (<1 m spacing) continuous foundation is preferred. To close the gap between the piers, curtain walls of 30-50 cm height should be constructed. The depth of foundations should be in the range of 50-60 cm depth and should rest on undisturbed soil. If the area is susceptible to frost the depth of foundation should be below frost line. The materials used for foundation could be RCC, galvanized steel, treated wood and concrete masonry. The foundation wall should be continuous. A 150 mm wall is usually sufficient for building spans up to 7.5 m. Use a 200 mm wall for wider building spans. Fig. 11.2 shows the typical foundation of a green house.

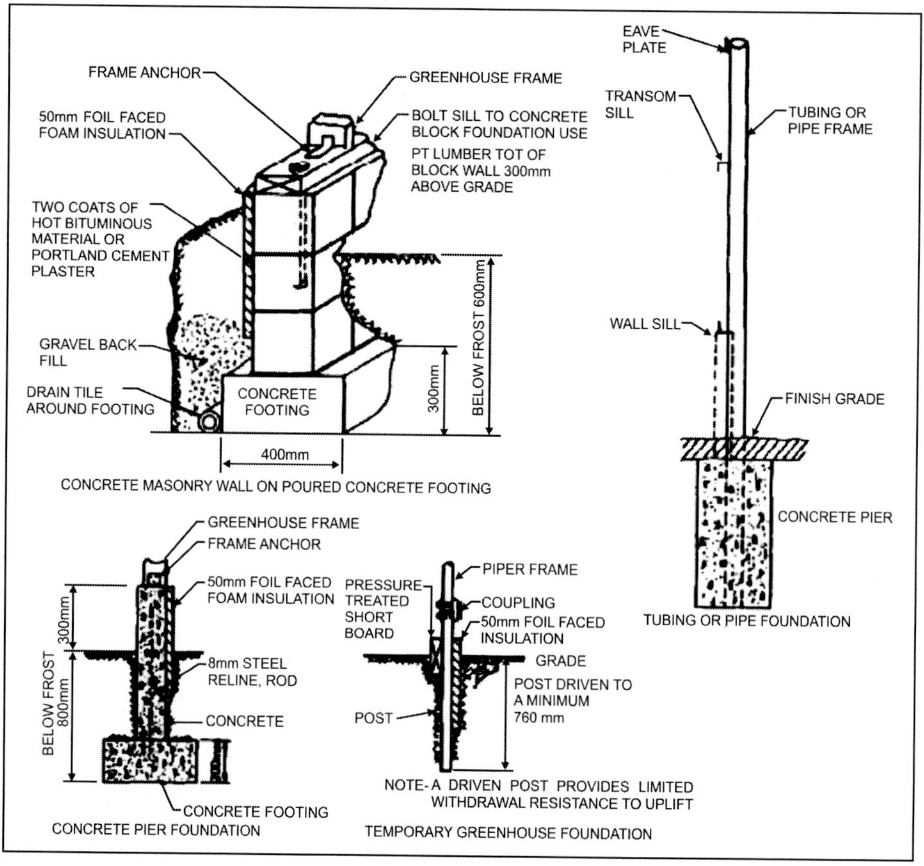

Fig. 11.2: Typical foundations for green house (as per IS standards)

11.5.2 Floors

Good floors are required in green house for drainage and weed control. The common materials used for flooring are gravel, pea stone, and trap rock. Porous concrete floors of 50 to 75 mm thickness have been found satisfactory for green houses. This is made from uniform sized aggregate of 10 mm size and cements water. Aisles and heavy traffic areas should be concrete. Thickness depends on the traffic load but usually 75 to 100 mm is sufficient. Concrete walks should have a broom finish for safety. Floors should slope to assure surface drainage and be sufficiently even to prevent puddling.

11.5.3 Frames

The most commonly used frames for green houses are wood, steel, aluminium, and reinforced concrete. If wood is used, it should be treated to protect it from termites, toxic effect of fertilizers and moisture. The treatment can be done using (coal tar, chromate copper arsenate (CCA) and ammoniacal copper arsenate (ACA)). White paint on either material will improve the light reflection in a greenhouse structure. The rate of heat loss through steel or aluminium is much higher than through wood, so metal frames may need special insulation. To avoid such heat loss through steel or aluminium, composite materials are sometimes used, such as a trussed beam of wood and steel or a member made of fibreglass reinforced plastic may be used.

11.6 Structural components of a greenhouse

11.6.1 Hoops

Hoops are the integral part of a greenhouse. These are semi-circular shaped pipes formed of bending galvanized iron pipes (15 mm diameter and 5.9 m long). The hoops provide support for placing and securing the covering materials. These materials should be strong and light weight

11.6.2 Foundation pipes

Foundation pipes are used to provide support to the hoops and to secure the polygrip assembly firmly. The materials used for foundation pipes are 25 mm GI pipes (Class A) having a length of 90 cm. At its one end a 10 cm piece of mild steel plate of 25 mm length and 3 mm width is welded to one side of the foundation pipes. 9 number of holes of 8 mm diameter are drilled into it at 10 cm spacing. The flat welded end is fixed to the ground to a depth of 75 cm depth by grouting with concrete. 15 cm pipe is kept above the ground level and hold hoops and polygrip assembly. The foundation pipes are spaced at 1 m interval.

11.6.3 End frame

End frames are provided on both ends of green house for supporting doors, supporting fixtures for environmental control equipment. The frames can be of wood, pipe frames or trusses. Truss frames are best suited for pre-fabricated structures.

11.6.4 Lateral support

Lateral supports are provided structural rigidity of end frames of green houses. These are generally fabricated from 10 mm diameter mild steel rods.

11.6.5 Polygrip assembly

Polygrip assembly is used to firmly secure the polythene covering material with the foundation pipe to protect from blowing away by wind. The polygrip assembly is generally made up of 20 guage galvanized iron strips. It is designed in such a way that it don't tear the polythene material.

11.7 Cladding Materials and their Properties

The cladding materials of a green house must be very transparent to PAR but at the same time should be good thermal insulator. The cladding materials must be able to filter the UV rays and infrared rays. One of the most important properties of cladding materials is to transform direct light to diffused light which enables maximizing the total effective radiation into the green house. Cladding materials should be environment friendly in production and use. The materials should also prevent condensation at the inner side during winter. Dust particles should slip over the house so that accumulated dust particles don't reduce the transmittance. Durability, ease of handling and cost of the materials are few other points one considers while selecting the materials. The table 11.2 gives an idea of suitability of different cladding materials.

Table 11.2: Properties of cladding materials

Materials	Trade Name	Transmittance to PAR (%)	Thermal Transmittance (%)	Durability (year)	Remark
Glass	Double strength	89	3	25+	Breakable, heavy but very transparent
	Insulated	71	<3	25+	
	Low Iron	90-92	<3	25+	
Acrylic	Plexiglass	93	<5	20+20+	Light weight but costly
	Double Wall	83	<3		

Materials	Trade Name	Transmittance to PAR (%)	Thermal Transmittance (%)	Durability (year)	Remark
Poly carbonate	Single wall	87-92	<3	7-10	Higher expansion
	Double wall	80	<3	7-10	
Fibre Glass	Filon	88	<3	10-15	Low cost
Polyethylene	Sun selector	85-90	<20	2-3	Short life span
Polyester film	Mylar	85-88	30	7-10	Small width
PVF	—	92	21	>10	Costly
PVC Film	—	80	10	4	Very strong
Woven PE	—	78	20	4	Very strong

The commonly used cladding materials are discussed below:

11.7.1 Glass

Glass is one of the popular materials used for covering a green house. Improved glasses are available with high strength and pane size. Although cost of glass is high but with new technologies cost have been reduced significantly. Glass provides good transmission of light over a period of time. However, maintenance is required to keep it clean to maintain the level of transmittance. Glass can transmit 84-87% of available visible light. However, in practical 70-75% is transmitted due to frames and other supporting structures. Double and triple glazing may be used to reduce heat loss.

11.7.2 Acrylic

The acrylic is available as thick or thin wall double extruded sheets. This provides some additional thermal insulation. These sheets are available in 1.2 × 1.2 m sizes and several thickness may be used. It has long service life of 20 years and more. The light transmissivity is upto 81%. It is having moderate impact resistance but can be easily scratched. Another advantage of the material is it has high expansion and contraction coefficient. However, the acrylic material is inflammable and expensive.

11.7.3 Polycarbonate

The polycarbonate material is similar to acrylic sheets. These sheets are available is size up to 2.4 m. These materials have long life up to 15 years. It has good life up to 15 years, but within 8 to 10 years the transmission capacity reduces and becomes yellowish. To overcome the yellowing and transmission of light of the sheets coated or treated materials is recommended. The material transmits light up to 77%. It has excellent resistance to impact and flame spread.

11.7.4 Fibre glass reinforced plastic panels (FRP)

FRP consists of polyester resins, glass fibres and can have minor amounts of other resins. These sheets are available in large sizes and different thickness. The average life f these sheets range from 5 to 12 years but treatment in every 5 to 6 years can enhance its life to retain its light transmissivity. FPR material transmits around 82 % of visible light. These materials have also high impact resistance and flame spread. These materials are corrugated and thus causes heat loss due to large surface area.

11.7.5 Polythene

Polythene is another widely used material due to its low cost and strength. It has high transmissivity of light (around 90%). It is available in wide sheets and have an average life of 3 years. It is affected by ultra violet light present in the sun rays.

11.7.6 Polyvinyl Fluoride film

This is a long-life film with high light transmission and strength. This film may be placed over aluminium extruded or steel frames.

Ethylene Vinyl Acetate (EVA) is another copolymer containing 14-18% Vinyl Acetate which can be used as alternative to polyethylene but this material suffers from stretching. In order to avoid the problem of expansion or stressing and excessive rigidity of the materials, co-extrusion technology is also available. Co-extrusion technology enhances the properties of the materials through combination of good properties of different polymers. Co-extrusion technology gives the materials durability, flexibility and UV stability. Anti-fogging additive can also be given in the co-polymers.

Copolymers have become very popular among the commercially available cladding materials due to low cost and high flexibility and easy to handle. Though the durability is not good, farmers find easy to invest smaller amount of money for such construction.

The recent addition of cladding material is TefzelT[2] film (Ethylene Tetrafluroethylene). This film was earlier used as transparent covering for solar collectors. The anticipated life expectancy is 20 years or more. Light transmission is about 95% and more than that of nay other cladding materials. A double layer has a light transmission of about 90%. But the disadvantage of this film is that it is also transparent to IR radiation and as a result less heat is trapped inside the green house.

11.8 Electrical requirements

Electricity is required in greenhouse for providing artificial lighting, operating heating and cooling system, for running the water pump system. In case of power failure alternate power arrangements have to be kept ready. Minimum power requirement for different sizes greenhouses are given below in Table 11.3.

Table 11.3: Electrical Power Requirement in greenhouse of different sizes

S.No	Green house size (m^2)	Amp/volts	Electrical power requirement (KW)
1	500	60/240	15
2	500-2000	100/240	24
3	2000-3000	150/240	36
4	3000-4000	200/240	48
5	4000-8000	400/240	96
6	8000-12000	600/240	145

(Ponce *et al.*, 2014)

11.9 Water requirements

Water is an important input in a green house. It depends on the cultivation medium, area to be irrigated, crops grown and heating and cooling system. The average water requirement during peak summer is around 2000 lit per 100 m^2 per watering. As per IS standards the water requirement for different sizes greenhouse are shown in Table 11.4. The water to be used in green house should have pH in the range of 6.5 to 7.5. The water should be applied in the green house by micro irrigation system. The quality of water should comply to IS 11624.

Table 11.4: Estimated maximum daily requirement

Crops	Water required (Lit/m^2)
Bench crops	16.0
Bedding plants	20.0
Pot plants	20.0
Chyanthenum	41.0
Roses, tomatoes	29.0

11.10 Instrumentation

Maintenance of appropriate climate within the green house is important to obtain satisfactory crop performance. So instrumentation is a important component in green house production. Instrumentation is necessary to monitor parameters such as temperature, humidity, carbon dioxide concentration, light levels etc within a greenhouse structures. With the availability of technology, it is possible to monitor these parameters of micro climate and plant interactions. Different parameters are discussed below

11.10.1 Air Temperature Measurement

Air temperature can be measured with the help of thermometers. The temperature measurement is affected by solar radiation so it is necessary to take precaution for accurate measurements. Air velocity have important role in getting accurate measurement. The recommended air velocity for accurate air measurement is 3m/s. The most commonly used temperature sensors used in green houses are either thermocouple or thermistors.

Thermocouples: A thermocouple is a very simple device consists of two wires of different materials held together. It is a most common device used to measure air temperature. It consists of thermo electric sensors known as thermocouples. It operates on the principle of thermo electric motive force (emf) generated in the device when the ends are maintained at different temperature. The magnitude of the emf being related to temperature difference. Mili-voltmeters are incorporated in the circuit to measure the emf generated into the circuit. A thermocouple consists of two different metals that are combined (usually by twisting them) to produce a voltage related to their temperature difference. There are several types of thermocouples that contain different metals, wire thicknesses, and thus have different costs and degrees of accuracy. Most commonly used materials is copper- constantan (constantan is an alloy consisting of copper (50-60%) and nickel (40-50%). The material has linear characteristics and withstands temperature at which it is used. Mercury thermometers are inadequate for most commercial production situations; they are hazardous if broken, don't allow automated measurements and lack precision. Thermocouples need to be calibrated before use.

Thermistors: Thermistors are thermally sensitive variable resisters made of ceramic like semi conducting materials. It is useful for continuous dynamic measurement. The advantages and disadvantage related to thermistors are:

Advantage

- It is very sensitive to temperature fluctuation. If properly calibrated the accuracy is +0.01 °C.
- It can withstand electrical and mechanical stresses.
- Thermistors have very wide range of temperature measurements in the range of -100 to +300 °C.
- The prices are affordable and easy adaptability to the available resistance bridge circuits and,
- High sensitivity and availability in extremely small sizes enabling a fast speed or thermal response.

Disadvantage

- Highly nonlinear resistance temperature coefficient.
- Problems of self-heating.

Infrared thermometer: Infrared thermometers are non-contact thermometers useful for measuring leaf temperature. Infrared thermometers typically have a conical field of view that depends on sensor make and model, and distance from the surface being measured.

11.10.2 Radiation Measuring Instrument

In greenhouse systems generally two types of light intensities are used for crop production. One of them measures the energy content (W/m2 or MJ/m2/day) and other is photon or quantum content (μmol/m2/s or mmol/m2/day). The plants generally used the photon in the range of 400-700 nm for photosynthesis purpose. The energy sensors measure solar radiation in the range of 300-2500 nm.

- **Net radiometer:** Measures the difference between the incoming and outgoing radiation in W m^{-2}. This sensor is typically used to evaluate the radiant energy environment in the greenhouse.
- **Pyranometer:** Pyranometers measures total hemispherical solar (beam + diffuse) radiation, usually on a horizontal surface. If shaded from the beam radiation by a shade rings or disc, a pyranometer measures diffuse radiation. The unit of total solar radiation is W m^{-2}. This sensor is typically used to evaluate the incoming radiant energy from the sun.
- **Quantum sensor:** Quantum sensors are used to measure PAR (photosynthetically active radiation) in μmol (photons) m^{-2} s^{-1} .

11.10.3 Photosynthesis Analyzer

Photosynthesis analyser measures the CO_2 concentration within the greenhouse system. Infrared gas analysers (IRGA) are most commonly used instrument for CO_2 concentration. It makes rapid, simultaneous measurements of photosynthesis rates and conductance with minimum change to the environment of the point material. With typical CO_2 depletions of only 2 to 100 ppm (depending on photosynthesis rate) and very small relative humidity change, the measurements are made very near to ambient conditions.

11.10.4 Humidity measurement

Humidity is a important parameters that affects the transpiration process of a plant. It is the amount of moisture present in water vapour. It is expressed as absolute or relative humidity.

- ✓ **Absolute Humidity:** Defined as mass of water vapour/ volume of gas and is usually expressed as g/m^3.
- ✓ **Relative Humidity:** It is defined as humidity of air with humidity of saturated air at same temperature and pressure. It is expressed as the ratio of mass of water vapour present in a given volume of air to the mass of water vapour necessary to saturate the same volume of air at same temperature.

Humidity can be measured by using dry bulb and wet bulb thermometer, hair hygrometer or humisters.

Dry bulb and wet bulb thermometer: Traditionally the wet bulb and dry bulb temperature is measured and is converted into relative humidity from psychrometric chart.

Hair hygrometer: Hair hygrometers consist of materials highly sensitive to moisture. These materials can be human hair, animal membrane; wood and paper undergo changes in linear dimension when they absorb moisture from atmosphere. Hair hygrometers can measure relative humidity in the range of 40-95% in the temperature range of -5 to 35 °C

Humistor: Humistor is the humidity sensitive resistor. The humistor consists of two metal grids bonded to a sheet of plastic. The arrangement is a given coating of moisture sensitive materials, i.e. hygroscopic slats such as lithium chloride, barium fluoride or aluminium oxide. As the RH rises, the film becomes more conductive and the electrical resistance of the grid is lowered. They can measure humidity in a wide range (0-100%) and is fact. The instrument gives low accuracy at very low humidity (< 5%).

11.10.5 Leaf Area Index (LAI) Measurements

The amount of foliage and its distribution is fundamental to radiation penetration. Thus it governs leaf temperature, turbulent transport, productivity, evapo transpiration, precipitation interception and soil temperature.

11.11 Errors and Uncertainties in measurement micro climate within the green house

During measurements different types of errors may be encountered. These errors can be classified as

(a) Systematic or cumulative error

  i. Instrumentation error

  ii. Environmental error

  iii. Loading error

(b) Accidental or random error

i. Inconsistencies associated with accurate measurements of small quantities

ii. Presence of certain system defects

iii. Effect of uncertain and randomly varying parameters.

(c) Miscellaneous type gross errors

(i) Personal or human errors

(ii) Errors due to faulty components/ adjustments

(iii) Improper application of the instruments.

Chapter - 12

Hydroponics: Soil less Cultivation

12.1 Introduction

Plants need sunlight, water and CO_2 and essential nutrients for its growth and development as well as production. Soil gives an environment where all these can be available. Soil is called as living body due to the presence of billions of micro-organisms which help to make soil fertile. If we take away all of these from soil, it becomes simply a container and not capable of supporting plants except providing the anchorage. In other words, if it is possible to supply all the necessary nutrients along with water, air and sunlight in adequate quantity, soil is not important for crop production. The Hanging Garden of Babylon and Floating garden of Aztecs in Mexico are the examples of such soil less farming and early instances of hydroponics.

The word 'Hydroponics' is derived from two Greeks words: *hydro* means water and *ponics* means labour, together means 'working water'. By definition, it is the science of growing plants in nutrient rich solution or in moist inert materials instead of soil. It is also called soil less culture. In hydroponics, plants are primarily grown either in nutrient solutions or in nutrient rich water with or without mechanical support from inert materials such as sand, gravel, etc. Some other terms are also used to refer hydroponics such as hydro-culture, nutri-culture, chemical culture, tank farming, aquaculture, etc.

Though hydroponics has the historical reference to the ancient civilization, the modern hydroponics got some amount of scientific references in the nineteenth century. During World War-II, US army has gone for extensive vegetable cultivation through hydroponics to supply fresh vegetables to its army in the Western Pacific Islands. Since, 1980s, hydroponics gained commercial status by producing vegetables and flowers. By 1995, about 60,000 acres of area was under hydroponics in green houses. Cucumber, Tomato and Pepper are the major crops grown commercial through hydroponics. About 55,000 acres area is under hydroponics in US, Canada and Mexico.

Precision agriculture, automation and resource efficiencies are the buzz words in modern agricultural system. Scientists are search of technologies and techniques where efficiencies of the inputs can be enhanced. Hydroponics is such a culture, where the efficiency of water, fertilizers can be increased tremendously. It needs precise management of everything, including the sunlight. Hydroponics does not need large amount of water for growing crops which otherwise needed in irrigation. Since hydroponics provides the mechanisms to supply nutrient in precise quantity including the trace elements, the quality of fruits is also greatly enhanced. The advantages and disadvantages of hydroponics culture are given hereunder:

Advantages of Hydroponics

i. Crops can be grown at small places include terraces, balconies where soil is not available. Hydroponics is ideal for cities, metros where people can plan cultivating vegetable at home through this technique.

ii. Crops can be grown in the area with very poor soils or diseased soils.

iii. There is no problem of weed in this cultivation process.

iv. Large number of labours is not required which is otherwise necessary for tilling, cultivation, inter-culture in field conditions.

v. Fossil fuel consumption for conventional agricultural operation is eliminated.

vi. Maximum yield per unit area is possible through hydroponics.

vii. Highest nutrient and water use efficiencies can be achieved.

viii. Soil borne diseases can be easily eradicated.

ix. Complete control of environment including rhizosphere environment is possible in hydroponics.

x. Balance nutrient with trace element enhance the nutritive quality of the crops.

xi. Automation is another possibility in hydroponics.

xii. Amateur horticulture is another avenue in hydroponics.

Hydroponics has definitely all ingredients of efficient and highly productive agriculture but at the same time certain disadvantages that pull back the system from getting wide spread acceptance among the farmers.

Disadvantages of Hydroponics

i. The initial investment of hydroponics system is quite high.

ii. Hydroponics culture requires sufficient skills to make it a successful venture

iii. There is no single composition of nutrients in hydroponics. Every crop needs specific nutrient solution and an extensive experimentation is needed to prepare nutrient solutions for every variety and species of crop. This often creates bottleneck for hydroponics farming.

iv. If the system is a closed one i.e. the nutrient solution is recycled in the system, any disease can spread very easily within the system.

v. Precise maintenance of the system is necessary to avoid the damage to crop due to nutrient stress or water stress. Daily attention is necessary unlike open cultivation.

12.2 Types of Hydroponics System

Depending upon the type of nutrient management and the growing media the hydroponics system can be of different types. It may be medium less i.e. only in nutrient solution or in medium where inert materials are used to grow the crops. Considering the mechanisms of the hydroponics systems, it can be either active systems or passive systems. Active systems have moving parts such as pump as an integral component. In passive system there is no moving part and the root system can take up nutrient from nutrient solution pools. We shall discuss commonly used hydroponics systems.

12.2.1 Standing aerated nutrient solution

This was one of the oldest hydroponics system and primarily used for research purposes to know the requirement of nutrients or the effect of elements in nutrient for crops. This is an active system where the single plant is supplied with measured doses of nutrients as per experimental design and the nutrient solution is continuously aerated with the help of pump (Fig. 12.1). This system is still used by the scientist for various plant nutrient studies. The requirement of this system is:

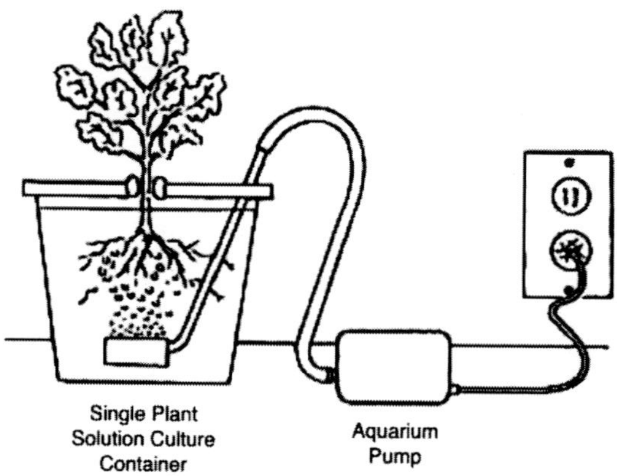

Fig. 12.1: Standing aerated nutrient solution system

a. One root vessel where plant is grown

b. Nutrient solution as required in the cultivation process

c. A pump and air tube for aerating the nutrient solution.

In this system, the nutrient solution requires periodic replacement, usually every 5 to 10 days, the frequency of replacement is dependent on the number of plants per vessel and the size of the vessel. Water loss from the nutrient solution will be replaced daily specially either with fresh water or $1/10^{th}$ strength of nutrient solution. However, there is a risk of creating imbalance in the nutrient concentration by this way of replacement hence periodic testing of nutrient quality is prescribed. This system has limited commercial use and also involves in lot of water requirement due to replacement of solution. This system if not used for specific nutrient experiment can use Hoagland solution of nutrient or variation of it as per requirement of the crops. The Hoagland solution is given in the Table 12.1.

Table 12.1: Composition of Hoagland solution for Hydroponics

Reagent	Concentration	
	Formula	**g/l**
Stock Concentrate #1		
Potassium Nitrate	KNO_3	50.55
Potassium phosphate	KH_2PO_4	27.22
Magnesium sulphate	$MgSO_4 . 7H_2O$	49.30
Micronutrient concentrate		100 ml
Micronutrient concentrate formulation		
Boric acid	H_3BO_3	2.850
Manganese sulphate	$MnSO_4 . H_2O$	1.538
Zinc sulphate	$ZnSO_4 . 7H_2O$	0.219
Copper sulphate	$CuSO_4 . 5H_2O$	0.078
Molibdic acid	$MoO_2 . 2H_2O$	0.020
Stock Concentrate #2		
Calcium Nitrate	$Ca(NO_3)_2 . 4H_2O$	118.0
Sequestrene 330 Fe		5.0
To use 1:200 dilution in water		

The final composition of the solution is given in Table 12.2

Table 12.2: Final chemical composition of Hoagland solution

Element	Quantity (mg/l)	Element	Quantity (mg/l)
Calcium	302	Manganese	0.974
Potassium	283	Boron	0.536
Magnesium	37.8	Zinc	0.300
Nitrate N	321	Copper	0.076
Ammonium N	39	Molybdenum	0.155
Chlorine	65	Sodium	4.56
Phosphorus	2	HEDTA	13.0
Iron	2.76		

12.2.2 Nutrient Film Technique (NFT)

This is the purist form of hydroponics. In NFT method the roots of the plants remain suspended in a trough (furrow or channel) through which nutrient solution passes. This is an active method without any solid medium for support. The plants are supported by specially made holder. Nutrient solution passes continuously and the root systems of the plant can extract the necessary nutrients from the solution. The furrows, individual or an array are inclined at about 1^0. The nutrient solution is released at the top end and allowed to pass through which is then collected at the bottom end where the nutrient reservoir is kept. The spent solution which is collected at the reservoir is then re-circulated. In is necessary to test the nutrient status intermittently as the nutrient will be absorbed by the plants. The nutrient content of the spent solution after few passes is reduced; enrichment of the solution becomes mandatory else the plant will be adversely affected. One needs to regularly monitor also the pH and Electrical Conductivity (EC) of the solution used for NFT hydroponics. Problem only occurs when the root density increases and put obstruction in the channel for the nutrient solution and the lower end plants may get malnourished.

Most troughs in use today are made of various plastic materials, the requirements being opacity, structural strength, and ultraviolet (UV) resistance. The design of the trough (width, height, and form) is usually determined by the crop to be grown. Lack of structural strength can lead to unevenness in the trough bottom that allows nutrient solution to lie in depressions that can lead to anaerobic conditions. The plants are set in the trough at the spacing recommended for that crop. Usually, plants are started in germination cubes made of fiber-glass or similar material. The cube with its started plant is set directly in the trough. Experience has shown that the germination cube should not be made of materials that disintegrate with time. A durable germination cube helps keep the plant set in place in the NFT trough. Normally NFT systems are closed systems, that is, the nutrient solution exiting the end of the trough is recovered for reuse. The addition of make-up water, the need for reconstituting the pH and nutrient element content, filtering, and sterilization are procedures that need to be established. An open system would mean that the nutrient solution exiting the trough is discarded, which is costly in terms of water and reagent use as well as posing a problem for proper disposal.

The major advantage of this system is that it is relatively cheap and easy to manage particularly the closed system. If the NFT practitioners have the facilities to control the pH, EC and other nutrient measurement, this system may prove to be good and profitable. A schematic diagram of NFT system is given in Fig.12.2.

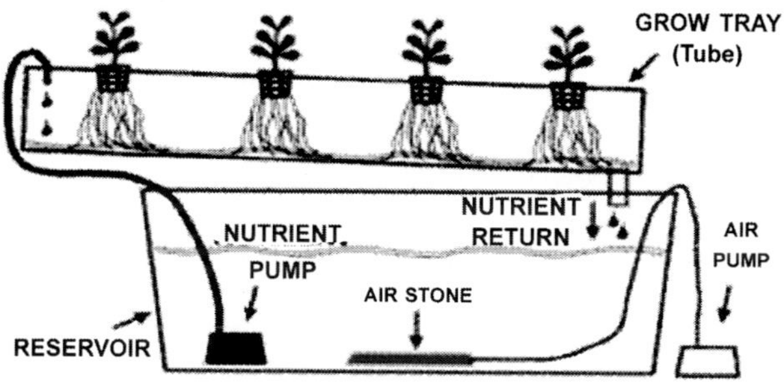

Fig. 12.2: Nutrient Film Technique of Hydroponics

There can be variation of this system depending on the types of trough used. Now-a-days, PVC pipes are also used to replace the trough. The plants are grown put in the holes made in the pipes at the recommended spacing. Nutrient solution is then pumped into the pipe which is connected to nutrient reservoir at both ends. The quality of the nutrient solution is continuously monitored in the reservoir.

12.2.3 Ebb and Flow Nutrient Solution System

This method requires inert growing medium to hold the plants to give mechanical support. If the growing medium has high porosity and water holding capacity, it can also hold water for some time for continuous supply of nutrient to the plants. This system is also an active system. It is based on the concept of intermittent flooding and draining the root zones. Each time flooding is done the nutrient solution washes the roots and as the solution goes back to the reservoir, the root can get fresh oxygen. This system needs continuous power supply to run. Power failure will damage the crops. A schematic diagram of Ebb and Flow system is given in Fig.12.3

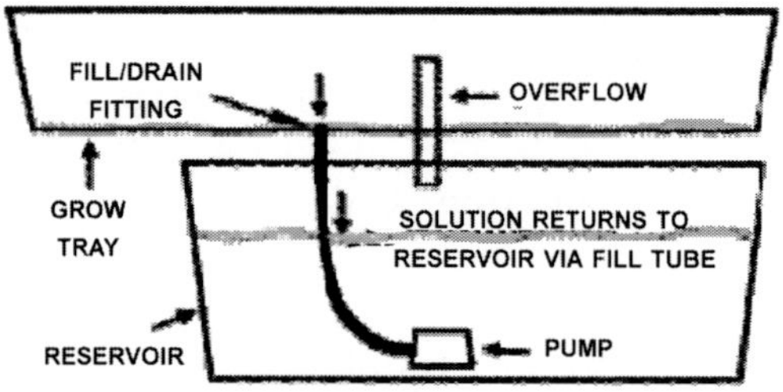

Fig. 12.3: Ebb and Flow System

12.2.4 Capillary System

Capillary system is a system with growing media, where the plants are grown in inert medium kept in a deep tray. This is a passive system where no moving part is available. However, if desired an air pump can be added to aerate the nutrient solution in the reservoir. The growing media for this system must have capillary force or metric potential similar to soils. Wicks are fitted at the bottom of the media tray the other ends of which are dipped in nutrient solutions. The wicks take up the nutrient solution from the reservoir through capillary action and supply the nutrients to the roots of the plants. A schematic diagram of capillary system is given in Fig.12. 4.

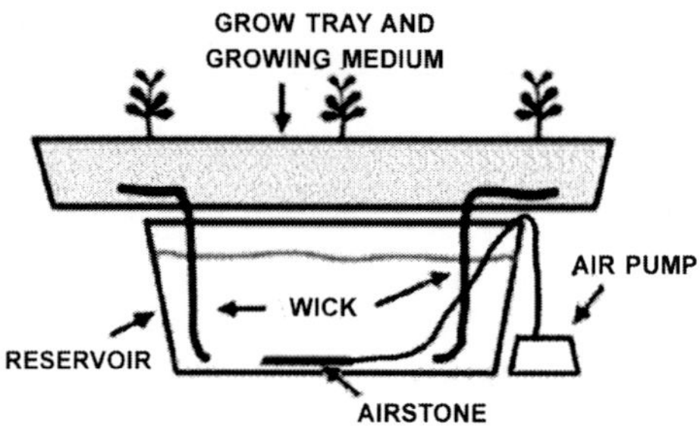

Fig. 12.4: Capillary System

12.2.5 Raft System

Raft System is very good for producing crops like lettuce quickly and in large quantity. In this system plants are grown in cups put in a Styrofoam sheets that floats on nutrient bath. High level of nutrient solution quality needs to be maintained in this system since the solution is stagnant in the bath. Therefore, continuous aeration is done through oxygenator.

12.2.6 Drip System

In drip system of hydroponics, the plants are grown in inert medium like rock wool, perilite, coco-peat, etc. and the nutrient solution is supplied with drip system. This system can be made automatic with timer as well as other sensors to keep the nutrient quality at its best and supply the water and nutrient at exact quantity as required. Crops like tomato, cucumber, pepper, etc. can be grown in large quantity under poly houses with this system.

12.3 Growing Media

The first criterion of hydroponics growing media is that it should be inert and non reactive to the chemical supplied. Modern day hydroponics growing media are far better than the river sand and gravel which were the early materials for soil less cultivation. The modern materials are capable of holding water and air in balanced quantity with their interstices. Fine sand or similar materials have very small interstices as a result it cannot hold sufficient air and water but the capillary action of find sand can be better than the coarse sand whereas the coarse sand has larger interstices which allow more spaces for water and air. If fast nutrient solution circulation is needed one can go for coarse medium like gravel, otherwise gravel cannot be considered as good hydroponics medium.

12.3.1 Coco-peat

It is an organic medium made of shredded coconut husk. Commercially the coco coir is called coco-peat. It is also commercially available as Ultrapeat, Coco-teck, etc. It has perfect balance of water retention and air spaces. It can also be locally available. Finely shredded and steam sterilized coconut coir offer a perfect medium free from any fungus and disease. Unlike peat moss, coconut coir does not decompose or deplete easily can renewable. Most commonly available cocopeat is in briquette form. The cocopeat is expandable on hydration and up to six times of volume is increased when soaked.

12.3.2 Perlite

Perlite is an amorphous volcanic glass that has relatively high water content, typically formed by the hydration of obsidian. It occurs naturally and has the unusual property of greatly expanding when heated sufficiently. It is an industrial mineral and a commercial product useful for its low density after processing. Perlite has excellent oxygen retention capacity. It is also used as soil amendment to improve soil permeability to avoid compaction. The main drawback of perlite is its low density which makes it unsuitable for flowing or flushing types of hydroponics as the materials will go away with water. Perlite is also great for rooting cuttings and fosters much stronger root formation than those grown in water alone.

12.3.3 Vermiculite

Perlite has high porosity and good aeration but passes water easily through it but vermiculite offers good water retention capacity which makes it suitable for hydroponics as a growing medium. Vermiculite is the name of a group of hydrated laminar minerals (aluminum-iron magnesium silicates) which look like mica. Horticultural vermiculite is processed with massive heat that expands it into accordion shaped pellets composed of multiple layers of thin plates. It does not rot, deteriorate, or mold and is enduring, odorless, non-toxic and sterile.

Vermiculite can be found in potting soil or purchased by itself in four different sizes for gardening with vermiculite. Smallest sized vermiculite is used for seed germination and as a growing medium and the largest size is used for improving soil aeration. Vermiculite is generally a neutral 7.0 pH, but is dependent upon the source from around the globe and its reaction is alkaline. It is very lightweight and mixes easily with other mediums. Vermiculite increases water and nutrient retention and aerates the soil, resulting in healthier, more robust plants. Vermiculite, although is less aerating than perlite, it is used as amendment for water-loving plants. Vermiculite is also used for soil conditioning and lightening either alone or in conjunction with peat or compost which accelerates the growth and promote anchorage for tender young root systems. Using vermiculite as growing medium will also enable the plant to more easily absorb the ammonium, potassium, calcium and magnesium necessary for vigorous growth. Vermiculite allows seeds to germinate more rapidly. Damping off is thwarted since vermiculite is sterile and the seedlings are easily removed without damage to the roots. The sponge-like quality of the vermiculite will absorb any excess moisture and prevent rot or mildew while protecting the plants from temperature fluxes.

12.3.4 Lightweight Expanded Clay Aggregate (LECA)

LECA (Lightweight Expanded Clay Aggregate) is an extremely coarse growing medium commercially available as Geolite, Grorox or Hydroton. It is expanded clay pellets capable of holding waer due its porosity and surface area. LECA are pH neutral and reusable making them very suitable for hydroponics.

12.3.5 Rockwool

Rockwool is also called mineral wool. It is made from molten rock which is spun into long fibres. These fibers are compressed as brick or block that can be used as growing medium in hydroponics. It can readily absorb water and also has very good drainage properties. The main benefit of Rockwool is its sterility due to which no pathogen can affect the hydroponic system. The fibrous nature Rockwool provides a good mechanical structure to hold the plant stable. The naturally high pH of mineral wool makes them initially unsuitable to plant growth and requires "conditioning" to produce wool with an appropriate, stable pH.

12.3.6 Sand and Gravel Culture

Sand and gravel culture though originally used as hydroponics medium and still can be used but they are considered to be poor hydroponics medium due to small interstices. Smaller interstices proved to be detrimental for good aeration and at the same time they are not good water holder. Any hydroponics system essentially needs both the properties. However, sand and gravel culture can be used for system like Ebb and Flow or in the Rat system where the growing cups can be filled with sand and gravel culture.

12.4 Nutrient Management in Hydroponics

Most of the plants grow in the environment of wide range of nutrient concentrations. From the practical stand points, hydroponics nutrient solutions, though crop specific, can have various combinations and growers can prepare their own hydroponics nutrient solutions themselves instead of purchasing readymade solutions for crops. All crops need macro and micro nutrients. The elements needed in higher quantities are called macro-nutrients. Nitrogen, Phosphorus, Potassium, Magnesium, Calcium and sulphur are needed in higher quantities whereas all other elements viz. Iron, manganese, zinc, boron, copper, molybdenum, chloride and nickel are required in smaller quantities. All the elements are supplied to hydroponics; however, nickel and chloride are not given in most of the hydroponics solution.

In hydroponics, it is absolutely necessary to start with testing of water for alkalinity, salinity and concentration of other salts in water. Alkalinity is measured in terms of calcium carbonate equivalents. Alkalinity values may range from 0 to 300 ppm of $CaCO_3$. Though pH is the measure of acidity or alkalinity of solution, $CaCO_3$ equivalence is more important than one shot pH value. With the level of alkalinity, one can decide about fertilizes for nitrogen. More acidic fertilizers can be selected for alkaline water to neutralize the alkalinity. As far as possible the pH should be about neutral however, a pH of 5.5 to 6.5 is workable. EC is a measure of total dissolved slats including both of essential elements and unwanted elements like sodium. Therefore, EC is a rough measure of water source purity. Knowing water source EC will help you determine whether to use an open or closed irrigation system. In closed hydroponic systems, the irrigation water is captured and reused, whereas in open systems, it's not. EC should ideally be less than 0.25 mS/cm for closed systems.In open systems, the build-up of salts can be managed by applying excess water to leach out soluble salts. Therefore, source water EC can be higher than in closed systems—ideally less than 1.0 dS/cm. Generally, most water will have dissolved sodium and chloride. Less than 50 and 70 ppm of sodium and chloride, respectively are acceptable.

Different crops require different nutrient depending upon its phenology and genetic capacity. Crops need nutrient at different growth stages too. Vegetative crops may require adjustment of different composition at different stages, but the fruiting crops may need tweaking of nutrients when it shift from vegetative to reproductive stage. It is also important to note that most hydroponic culture needs maintaining of two to three fertilizer tanks to avoid the problems of precipitation of sludge formation due to mixing of different chemicals together. For example calcium can form sulphate or phosphate when the concentrate mixes with each other rather than formation of ready to use mixture solution. They can be mixed in proper concentration separately and the final mixing is done at the time of application. It is also important to pick up quality chemicals which have 100% solubility in pure water. The standard nutrient solution for leafy green vegetables and tomatoes are given in the Table 12.3 and Table 12.4.

Table 12.3: Hydroponic recipe of leafy green vegetables

Element	Jack's Hydro-FeEd (16-4-17) (ppm)	Jack's Hydroponic (5-12-26) + $CaNO_3$ (ppm)
Nitrogen	150	150
Phosphorus	16	39
Potassium	132	162
Calcium	38	139
Magnesium	14	47
Iron	2.1	2.3
Manganese	0.47	0.38
Zinc	0.49	0.11
Boron	0.21	0.38
Copper	0.131	0.113
Molybdenum	0.075	0.075

Table 12.4: Tomato hydroponic recipe (in ppm)

Element	Weeks 0-6 Higher N, Ca and Mg for vegetative growth	Week 6-12 Lower N, higher K for reproductive growth	Week 12+ balance between vegetative and reproductive stages
Nitrogen	224	189	189
Phosphorus	47	47	39
Potassium	281	351	341
Calcium	212	190	170
Magnesium	65	60	48

Element	**Weeks 0-6 Higher N, Ca and Mg for vegetative growth**	**Week 6-12 Lower N, higher K for reproductive growth**	**Week 12+ balance between vegetative and reproductive stages**
Iron	2.0	2.0	2.0
Manganese	0.55	0.55	0.55
Zinc	0.33	0.33	0.33
Boron	0.28	0.28	0.28
Copper	0.05	0.05	0.05
Molybdenum	0.05	0.05	0.05

Chapter - 13

Crop Growth Models

13.1 Introduction

The growth and development of crops are comprised of complex photo-chemical and bio-chemical processes. The phenological characteristics of any crop, its growth pattern, duration of different growth stages and potential yield may depend largely on genetic potential; however, the actual productivity and the health of the crops are dependent on many stress factors which are largely environmental. Environmental factors like rainfall, temperature, sun shine hours, net solar radiation, relative humidity, soil types, nutrients, water, etc. all directly influence the crop growth behaviour at different stages and finally the actual economical yield. These environmental factors are truly dynamic in nature and change frequently at time scale, setting non-uniform growth pattern to the crops.

Crop growth factors can be classified into two groups, they are internal and external. Internal factors are genetic or hereditary. The following genetic characteristics are important for crop yield:

a. High yielding ability
b. Early maturity
c. Resistance to lodging
d. Drought and salinity tolerance
e. Tolerance to insect and pests
f. Chemical composition of grains such as oil content and protein content
g. Quality of grains such as fineness, coarseness
h. and quality of straw or stover.

The above characters are less influenced by environmental factors since they are governedby genetic make-up of crop.

However, about 50% of the crop yield is attributed to the climatic factors which can also be called as external factors. The external factors are:

i. Climatic
ii. Edaphic
iii. Biotic
iv. Physiographic and
v. Socio-economic

Growth process essentially consists of production of carbohydrate by the plant through photosynthesis and assimilation of it. Photosynthesis on the other hand consists of three processes *viz.*diffusion of carbon dioxide from atmosphere to the plant through stomata, photochemical process in the presence of sunlight and chlorophyll where synthesis of carbohydrate takes place and bio-chemical process where assimilation of carbohydrate takes place for actual physical growth of the plants.

13.2 Modelling Concept

A model is a representation of any system of the real world. Any model simulates the complex process that undergoes in any system. On the other hand, a system is an assemblage of interconnected objects or elements upon which any observation can be taken.Model may be of different kinds depending upon the way the system is represented. Broadly, model may be of four kinds:

1. Conceptual or verbal, where the model is described in natural language.
2. Diagrammatic: Diagrammatic models are described with the help of figures, drawings, flow-charts, etc.
3. Physical models, where a physical, scaled down version of the actual system is made.
4. Mathematical models are described through mathematical formulations. In mathematical model complex processes are simulated in such a way that the results of the processes in any system can be predicted without actually doing or going through it. Mathematical models can simulate complex bio-physical or bio-chemical processes through a set of equations of algebraic or differential functions. The mathematical models can be classified based on how the models are devised. If the models is formulated based on purely algebraic arguments usually differential equations, then the models are called analytical models, whereas if the models get the solutions through numerical approximation then they are called simulation models.

13.3 Types of Models

Models based on mechanisms involved

A model said to be **empirical, descriptive or phenomenological** if the models developed based on experimental output. Such models rarely explain the process involved. It doesn't explain the mechanisms to explain the cause of behaviour of the model. Regression models developed based on the recorded data of experiment fall under this category. These models usually consists of one or more mathematical equations. **Mechanistic or explanatory** models consist of quantitative description of the mechanisms and processes of the system. These explanations are statement of scientific theories or hypotheses. While formulating mechanistic model, one needs to analyse the process and then quantifies the mechanisms. Finally the model is built by integrating all the theories for the entire system.

Models based on involvement of time as variables:

> **Static model:** If the model doesn't have time as an explicit variable then the model is called static model. Static models are usually close to equilibrium process.
>
> **Dynamic model:** Dynamic models consist time as an explicit variable.

Models based on Involvement of Probability Functions:

> **Deterministic model:** Deterministic models yield definite values for any given inputs. For given values of initial and boundary conditions, a set of input values will always yield the exact values as output. In generation process of the model there is not probability function. However, deterministic models may also be linked to probabilistic model is the system as a whole.
>
> **Stochastic or Probabilistic Models:** Stochastic models contain random variables or some probability distributions within the model. If the uncertainty of variables is observed, stochastic model becomes necessity. The predicted quantity(s) of the models also follow the same probability distribution function as expected from the system. As for example, runoff model of watershed or rainfall prediction models are stochastic models as rainfall itself is a parameter of probability. These are some time series random events.

Definition of some important glossaries:

Parameters: A numerical measures of a property or characteristic that is constant for a system under specified conditions.

Coefficient: A variable or constant appearing in a mathematical model, each values of which defines the specific form of model.

State variable: A state variable is a quantity that defines the state of the system at a given point of time. The dynamic models based on the modelling problems to predict the state of the variables at any time.

Rate variables: These variables characterize the rate of change of each variables. The rate variables are the function of state variables.

Auxiliary variables: To improve the understanding of the process in the systems, some additional variables are included in dynamic models. These variable may not alter the principal process in the system but explain the process in a better way. Such variables are called auxiliary variables.

Driving variables: The variables specifying the environment and which vary autonomously with time are called driving variables. The driving variables affect the rate variables in the systems. The environmental variables such as temperature, humidity, wind speed, sunshine, etc. are called driving variables.

13.4 Stages of Model Formulation

Four stages of model development processes are recognised. These are conceptualization, formulation, programming and testing. Conceptualization covers the greater part of the model development process. This stage means the composite of all thought processes that gone through in analysing the problem and finding out the solution. Model conceptualization process can be considered as a stage of mixture of efforts of science and philosophy. Logical analyses and intellectual abstractions of the processes are taken place in this stage. Mathematical and numerical methodologies for representation the process is also defined here.

The second stage of model development is formulation. In formulation stage parametrisation and translation of the concept to the form of calculation is done. Formulation primarily means representing the process into some mathematical equations. However, formulation is also understood as preparation of algorithm or sequence of calculation. In formulation stage the algorithm may include 'look-up table', graphical relationship, finite different or finite element process or differential equations.

Programming is the third stage of model formulation. In this stage all the mathematical formulae and calculations are translated for effective calculations through programming language. After translation of the formulae into a programme, the model formulation enters into the final stage of testing. Testing means the examination of models, how the model is functioning or whether the model can predict the result for the given process. This involves in studying the goodness of performance through different statistical tools. The comparison between the observed and predicted values are carried out and accordingly the model can be accepted or reject or modified. Testing phase of models also includes sensitivity analyses of the model. The sensitivity study reveals the performance of the models against the changes of different input parameters or different variables (rate variables or driving variables). The steps of a crop growth model are given in the Fig.13.1

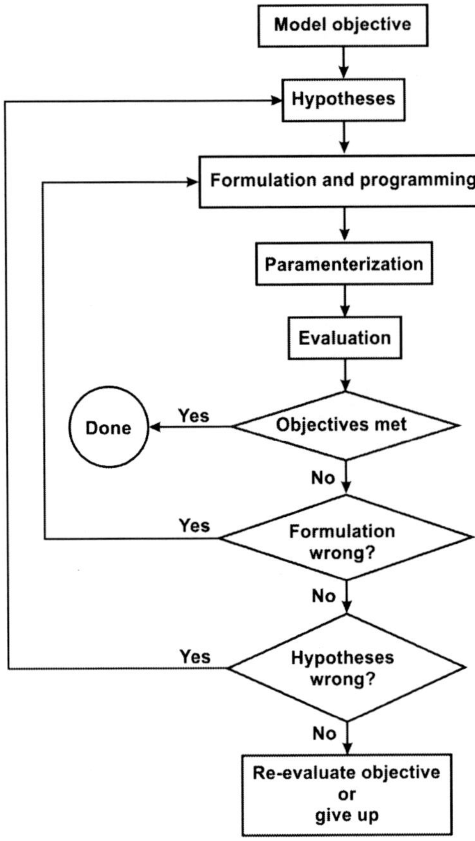

Fig. 13.1: Crop growth model development process

13.5 Forrester Diagram

Forrester's diagrams are designed to represent any dynamic system in which a measurable quantity flows between system components. The following are the conventions to represents the processes and components of any flow chart (Fig. 13.2) of system functions.

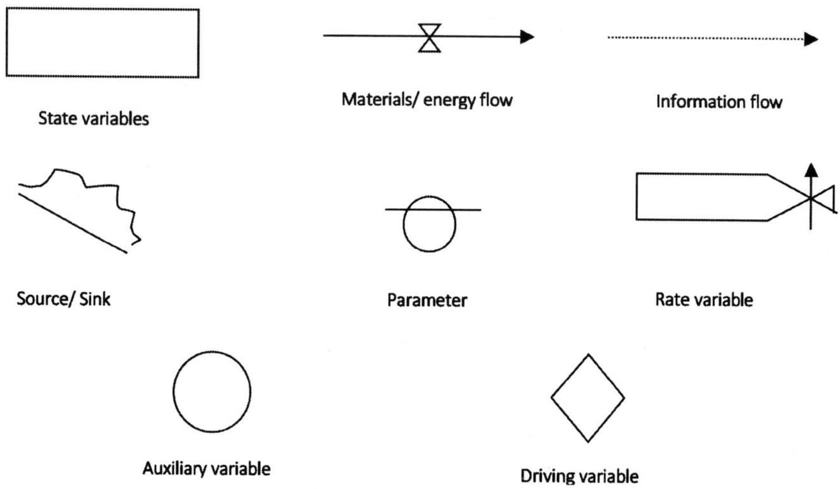

Fig. 13.2: Forrester's diagrams for representing parameters and processes.

Parameters and variables are already defined in Section 13.1. The materials or energy flow in any flow chart is represented by a bold arrow and identify the pathway over which the flow of energy or materials takes place. The rate of flow is represented by the control valve on the arrow and dependent on the process of the system. Information flow from one process to another process in the system is represented with a dotted arrow. The tail indicates the influencing component and the head of the arrow indicates the influenced component. Similarly, the inputs to the state variables and outputs from the state variables are represented by clouds. A complete relational diagram with these Forrester's diagrams is given in Fig.13.2

13.6 Importance of Crop Growth Models

As stated earlier, model mimics a system. It tries to emulate exactly the ways a system works. It is clear that a crop growth model is based on a set of equations or sub-models (sub-routine) which translate the complex bio-physical or bio-chemical processes that go in the growth and development of plants. Since models are based on all the mechanisms, singly or in combination, it often becomes easy to understand the influence of the individual processes how they affect on crop development and production. A crop model therefore enables us to find out answers to many questions in relation to the intricate process of crop production without going through extensive field experiments. How the genetic trait of crops can realise the potential yield, how temperatures affect the different stages of crop growth and vigour, how water stresses and fertilizer availability influence over all growth and productivity, is there any affect of climate change in the sustainability of crop production, etc. can be understood through crop growth models.

But one should realise that crop growth models are not the answer to all the questions. Models cannot substitute the field experiments. They can be at best supplementary. Models offer a simplified concept of real world. It is definitely a powerful tool to interpret the process and the influence of various factors involved in it. Crop models cannot be built without invoking a set of hypotheses and the simulation results cannot be rigorously tested without measurements that describe the performance of the crop over a wide range of environments. Hence to realise the true strength of models, filed experiments are always necessary. Nevertheless, crop models like other tools have strength and limitations. Therefore, the success of such tools depends on the quality of information of the parameters, strength of the models and proper understanding of the processes involved in the models.

Crop models have various applications. These are mostly grouped in three categories: research, management and teaching. Following are the areas where crop models can be effectively used:

Research:

- Integration of knowledge gathered through experiments
- Integration of knowledge across disciplines
- Improvement of experiment documentation
- Crop genetic improvement
- Crop management improvement
- Yield analyses
- Response to climate variability
- Environmental impact of crop production

Management:

- Development of best management practices through simulation
- Pre-sowing and mid course decision support
- Site specific precision management
- Pre-harvesting yield forecasting

Teaching:

- Student education
- Farmers education

Like any other tools, crop models also have both advantages and limitations. Depending upon the situtations these tools can be considered helpful or redundant. Some of the advantages are given:

a. Crop models considerably reduce the time of experiment in the field which may otherwise take long time or several season to gather desired information particularly in case of modfied or controlled conditions. It also becomes helpful for students working for degree programme due to their limited time availability within the curriculum.

b. Environments or sceanarios can be effectively created or altered in model environment for variable inputs to determine the simuated result.

c. Since simulated results can be obtained through model, the models often become very handy for contingency or distressed conditions for mid tem correction to overcome the distress for obtaining better output.

d. Many bio-physical processes cannot be effectively studied in laboratory conditions. These can be processed through models to have a greater insight of the process and can be presented in simpler form.

e. Any complex process in the system can be simplified and understood easily.

f. Climate change research and their adaptation can be very expensive if one creates the projected environment physically like the facilities for elevated carbon (FACE) or elevated temperature (FATE). Models can enable us to do the experiment with such climate scenario or even combinations of scenarios. If the models are calibtrated suitably for various parameters, they give us often very good results.

The crop growth models cannot be answer to all the complex issues and it should not be used as one. The limitation of the crops growth models are:

a. Like any models, if the models is not properly calibrated for the all the parameters, it may land one into a wrong conclusions.

b. Over reliance on crop growth models may lead to minimum exposure to field conditions, which anyway cannot be replaced by any level of modelling in true sense.

c. Real world phenomenon and its effect may be suppressed while working with the models. Students and researchers will be less exposed to real world problems which may be dynamic and are not incorporated while development of models.

d. It often becomes important to understand the conditions of model development. If one uses the model outside of the domain as the model is designed for, one may come out with erroneous result.

e. Due to software issues many a time operation of models becomes frustrating and one needs to update the models as well as the operating systems.

Crop models are undoubtedly good and alternative tools for many situations and hence they must be carefully used to avoid misleading conclusion. The models should be carefully used in teaching and research purposes carefully. The crop model should be complimentary tool and not to replace the real world experiment. The teachers should encourage the students to acquire cognitive advances in the process. The model users should be thoroughly exposed to the limitations, sensitivity of the models before actually using it.

13.7 Crop Growth Models

Crop productivity reaches its potential when ideal conditions can be given. The potential can be realised only when no limiting factors put brakes on growth process. Among the nutrients, nitrogen and phosphorus are two nutrients considered as growth and yield limiting if they are not supplied in optimum quantity. Similarly, water can be another growth limiting if crop comes under stress due to less supply. Crops models therefore distinguished four production levels by imposing these limitations. Production Level 1 is the condition where no constraint is imposed. All the inputs are at optimum level and even no biotic stress is considered. In case of production Level 2, growth is limited by availability of water. At production Level 3, the growth is limited by the availability of Nitrogen and whereas at production Level 4, the growth is limited by phosphorus. A relational diagram at production level 1 is given in Fig.13.3.

In this relational diagram, temperature and light are driving variables (in this diagram they are shown as underlined variable and not as diamond) which influence the rate variables photosynthesis and development. Photosynthesis efficiency is an external constant which also controls the photosynthesis process. Photosynthates then goes to the auxiliary variable which then influence the growth of the plants. Partitioning occurs depending on the stages of crop (state variable). The photosynthates go for the development of stem, branches and roots. Photosynthesis is also controlled by another auxiliary variable i.e. leaf area. Some parts of photosynthates are also utilized for maintenance and respiration. Growth rate depends on the amount of reserves available and conversion efficiency. The auxiliary variable, portioning, gets information from the development stage to determine the allocation of photosynthates to different organs. Shoot and root biomass are state variables which are represented as rectangles. This relational diagramme is valid only for production Level 1 where there is not limitation imposed and plant can yield to potential level as per its genetic capability.

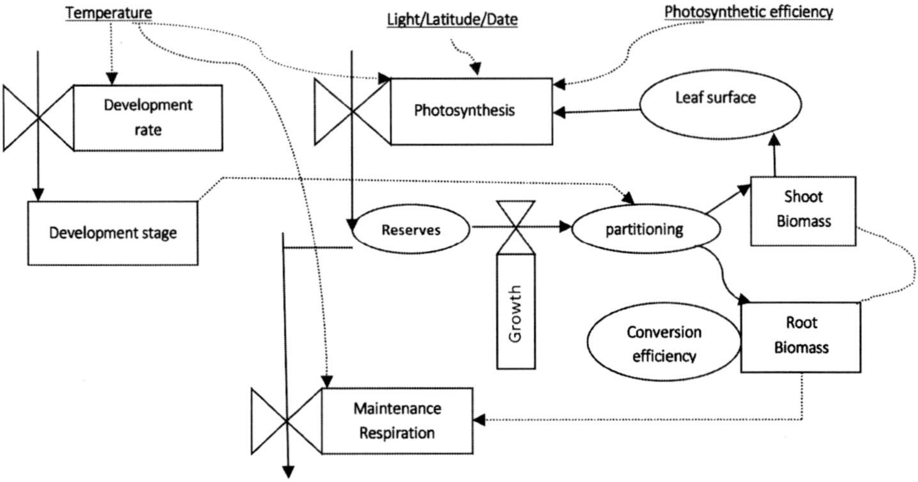

Fig. 13.3: Crop Growth model at production Level 1

13.8 Radiation Use Efficiency

Crop models determine the accumulated biological mass which is dependent on the incepted photosynthetically active radiation which is controlled by radiation use efficiency (RUE). Under non-limiting conditions (i.e. free from pests, diseases, and weeds, and experiencing non-limiting conditions with respect to temperature, water, and nutrients) the radiation use efficiency generally remains constant. The sugarcane model operates each day by inputting solar radiation data. Since solar radiation is available as total radiation, the first step in the model on each day is to convert the solar radiation data to a Photosynthetically Active Radiation (PAR) value. This conversion is readily done since PAR is about half of total Solar Radiation (SRAD) so the input radiation values are multiplied by 0.5. The daily amount of sugarcane growth is then simply calculated by multiplying the PAR value by sugarcane RUE. The RUE values for some important crops are given in Table 13.1.

The dry matter production in the crop growth process can be understood by the following simple steps.

PAR is calculated first as

$$PAR = 0.5 \times SRAD \qquad \text{...... (13.1)}$$

Daily Dry Matter Production is calculated as:

$$DDMP = PAR \times RUE\ (g\ m^{-2}\ day^{-1}) \qquad \text{...... (13.2)}$$

The total dry matter production (TDMP) can be calculated as:

$$TDMP = \sum_{i=1}^{n} DDMP_i \qquad \text{...... (13.3)}$$

Table 13.1: Radiation Use Efficiency (RUE) of some important crops

Crop	Radiation Use Efficiency (g MJ^{-1} PAR)
Rice	2.2
Wheat	2.2
Maize	3.5
Sorghum	3.5
Sugarcane	3.6
Cotton	1.8
Chickpea	1.8
Pea	1.8
Soybean	2.0
Peanut	2.0
Sunflower	2.1
Barley	2.1

13.9 Status of Crop Growth Models

The generic crop model SUCROS for potential production situation was developed, which formed the basis of Wageningen crop models such as WOFOST, MACROS, ARID CROP, SAHEL, PAPRAN, INTERCOM and ORYZA during the 1980s and 1990s (van Ittersum *et al.*, 2003). Spitters and Schapendonk (1990) developed a simple model (LINTUL) in which calculation of photosynthesis and respiration was replaced by radiation use efficiency. Later work by the Wageningen group resulted in development of ORYZA 2000 (Bouman *et al.*, 2001), GECROS and LINTUL3 (Shibu *et al.*, 2010). These modeling approaches have served as basis and inspiration for modeling groups around the world (van Ittersum *et al.*, 2003). Another well-known modelling group is DSSAT (Decision Support System for Agrotechnology Transfer), which was originally developed by an international network of scientists, cooperating in the International Benchmark Sites Network for Agrotechnology Transfer project (IBSNAT) to facilitate the application of crop models to agronomic research using a systems approach. Its initial development was motivated by a need to integrate knowledge about soil, climate, crops, and management for making better decisions about transferring production technology from one location to others where soils and climate differed (Jones and Kerswell, 2003).

DSSAT has been in use since 1989 by researchers worldwide. It incorporates models of about 30 different crops ranging from cereals to fibre crops with software that facilitates the evaluation and application of the crop models for different purposes. Recently, the DSSAT crop models have been redesigned and programmed to facilitate more efficient incorporation of new scientific advances, applications, documentation, and maintenance. Now, it is a collection of independent programs that operate together; crop simulation models are at its centre. A flow chart of DSSAT model is given in Fig.13.4.

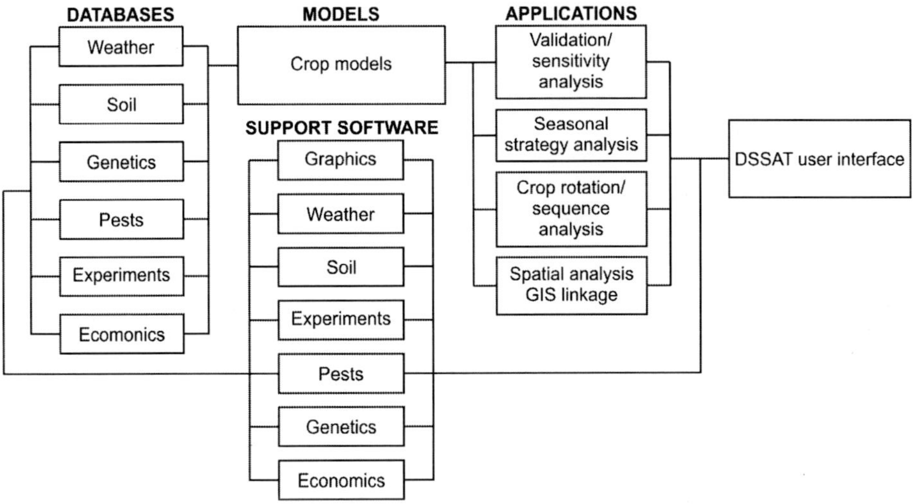

Fig. 13.4: General flow chart of DSSAT crop growth model

Chapter - 14

Climate change and its effect on Crop production

14.1 Introduction

Climate Change can be simply put as the change of global and regional climate pattern on a long time basis. It is definitely a defining moment of the present civilization which is threatening the very existence of the civilization due to shifting of weather, extreme events of rainfalls and drought, melting of glaciers and rise of sea levels. The impact is global and catastrophic. The climate change can be basically attributed to human fingerprints of greenhouse gases in every activity in the industrialized world. With the increased amount of greenhouse gases the global temperature is raising gradually which is influencing the rainfall pattern. The rise of temperature and the uncertain and extreme events of rainfalls are directly threatening the bio-diversity and agricultural productivity of the world.

Greenhouse gases are produced in natural processes and they are important to maintain earth temperature by creating a protective cover at the boundary of the atmosphere. They make the earth livable. However, due to rapid industrialization, large scale chemical agriculture, increase of vehicular population and burning of fossil fuels and deforestation resulted in increase of these gases and created the detrimental effects. There are some basic well-established scientific links:

- The concentration of green house gases in the atmosphere of the earth is directly responsible for global temperature rise;
- Since the industrial revolution, the concentration has been rising continuously;
- CO_2 accounts about two third of the total greenhouse gases and largely a product of burning of fossil fuels.

The Intergovernmental Panel on Climate Change (IPCC) had made some objective studies and gave some clarity on the climate change cause and effect. It has identified the fingerprint of human activities on the climate change as per its Fifth Report on Climate Change. Some of the important findings are:

- The average global temperature increased by 0.85 °C since 1880 to 2012.
- Due to melting of glaciers at a higher rate for rising temperatures, the sea level has increased. The global average sea level rose by 19 cm as oceans expanded due to warming and ice melted since 1901 to 2010.
- With current concentrations and at present rate of emissions, the temperature is going to increase to an alarming level. Oceans will expand and the ice cover is going to melt. Average sea level rise is predicted to be 24–30 cm by 2065 and 40–63 cm by 2100 relative to the reference period of 1986–2005. Most aspects of climate change will persist for many centuries, even if emissions are stopped.

Indian agriculture remains vulnerable to the vagaries of weather, and the looming threat of climate change may expose this vulnerability further. It is reported that climate change could reduce farm incomes by 15-18% in irrigated areas and by about 20-25% in unirrigated areas with potential reduction in yields by 4.5-9% in the short-run and by approximately 25% in the long-run in the absence of adaptation by farmers. In India, studies by several authors shown that during last century there is observed increasing trend in surface temperature, no significant trend in rainfall on all India basis and decreasing/increasing trends in rainfall in regional basis. Further, the number of extreme events has dramatically increased with less number of rainy days.

The most imminent climatic changes in recent times is the increase in the atmospheric temperatures due to increased levels of greenhouse gases such as Carbon dioxide (CO_2), Methane (CH_4), Ozone (O), Nitrous oxide (N_2O) and Chlorofluoro carbons (CFCs), Water vapour (H_20), etc. Because of the increasing concentrations of those radiative or greenhouse gases, there is much concern about future changes in our climate and direct or indirect effect on agriculture.

The climate change not only has its effects on crop and livestock produces but also on the economic consequences of these potential yield changes. Agriculture being a major human enterprise, the increasing atmospheric CO_2, rising temperature and altered precipitation pattern affects agricultural productivity reducing it significantly. Livestock production systems are at risk to temperature stresses, particularly the lengthened exposure to extreme temperatures will further enhance production cost and productivity losses associated with all animal products, e.g., meat, eggs and milk.

The effect of pest and diseases on the crops due to climate change is however not clear yet. Species migration in case of both flora and fauna is going to happen. While some of the organism may thrive well in the enhanced temperature some may change their habitat. Micro-organisms have strong adaptability to any change which is not evident on larger life and crops. The change may not pose any problems to the insects and pest but it will surely cause problems to crops. Weeds may often be found to assimilate CO_2 better and exhibiting vigorous characters or sometime behave as invasive in higher temperature situation. Extreme events have direct bearing on the natural resources. Heavy rainfall has the potential for flood and soil erosion. Erosion is however of particular concern. Changing climate will contribute to the erosivity from rainfall, snowmelt, and wind which ultimately affects the agricultural sector as a whole which conclusively affecting the Economy of the Country. Water scarcity due to drought will have devastating effect on crop production. Prolong drought will cause hydrological aberration to the agricultural eco-system. Rainfed agriculture is bound to face the brunt of the climate change. Plant response to climate change is governed by a complex set of interactions to CO_2, Temperature, Solar Radiation and Precipitation. Changes to the rate of soil organic matter accumulation will be affected by climate through soil temperature, soil water availability, and the amount of organic matter input from plants. The following figure (Fig. 14.1) explains the impact of climate change on agriculture and necessary interventions for mitigation and adaptations processes.

Mitigation of greenhouse gases in a given time frame as per Paris agreement is still not at sight due to lack of consensus among the countries especially among the countries responsible for maximum dumping of greenhouse gases. Even it happens; the impact of climate change on global agriculture is going to persist for a longer period. In order to offset the serious impact on agriculture, adaptation and adjustment are the only ways out. Development of drought tolerant, cold tolerant, heat tolerant varieties, adjustment of crop calendar with appropriate varieties to escape the period of possible of extreme events are some of the strategies. Other climate smart technologies available in the scientific domains may be useful against the climate change impacts.

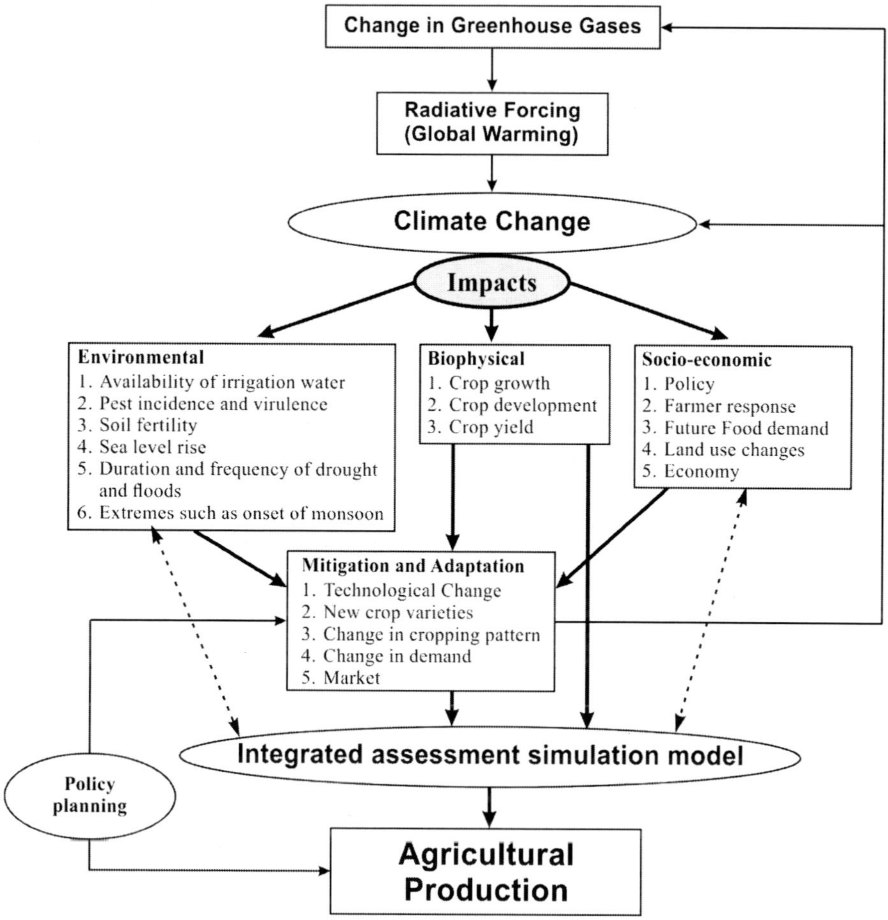

Fig. 14.1: Impact of climate change on agriculture

Climate models such as Global circulation Models (GCMs), regional Climate Models (RCMs) are coming handy for scientific community in simulating the effect of different possible climate parameters in a changes scenarios. The emission scenarios or Representative Concentration Pathways (RCPs) as prescribed by the IPCC are considered for radiative forcing to run these models. These give us the projected climate scenarios based on which any adaptive management can be possible to mitigate the impact of climate change in agriculture. Crop growth models also come handy in determining the impact of climate change on various crops and varieties. The projected changes of CO_2 concentration, rainfall and temperature can be easily incorporated in the Crop Growth Models after calibrating the models for certain crops and varieties and then estimate the effect. The crop growth models can also be linked to appropriate GCMs/ RCMs and the output of these climate models can be input for crop growth models to determine near possible scenario of agriculture. Such studies give sufficient scope of adaptation and modification of cultural practices and genetic resources for fighting with the ill effect of climate. Crop growth models further give enough scope to accommodate contingencies of extreme values and analyses to carry out mid course correction for aberration of climate during a crop season. The following sections are discussing the response of crops to climate change in Indian context.

14.2 Crop response to Climate Change

Plant response to climate change is dictated by a complex set of interactions to Temperature, CO2, Solar Radiation, and Precipitation.

14.2.1 Temperature

It has also been reported that Average annual temperatures have risen by around 0.48 degrees (between 1970 and 2016) and there has been a steady increase in temperature extremities. The number of 'very hot' days as well as the number of dry days has increased, consistent with models of climate change which predict increased variability in weather.

Pollination stage of the crops is one such critical stage that higher temperature can most adversely effect on crop performances. In pollination stage, pollen is released to fertilize the plant and trigger development of reproductive organs, for fruit, grain, or fibre. It is one of the most sensitive stages to temperatures, and exposure to high temperatures during this period can greatly reduce crop yields and increase the risk of total crop failure. Plants exposed to warm night temperatures during grain, fibre, or fruit production also experience lower productivity and reduced quality.

Higher temperatures tend to reduce crop yields and favour weed and pest proliferation. In one study, it was showed that there was a decline of 600–650 grains/m^2 in wheat crop with every 1^0C increase in mean temperatures above 17–17.7^0C during the terminal spikelet initiation to anthesis. It was estimated that a 2^0C increase in mean air temperature could decrease rice yield by about 0.75 t/ha in the high yield areas and by about 0.06 t/ha in the low yield coastal regions.

Each crop species has a given set of temperature thresholds that define the upper and lower boundaries for growth and reproduction, along with optimum temperatures for each developmental phase. Plants are currently grown in areas in which they are exposed to temperatures that match their threshold values. As temperatures increase over the coming century, shifts or transfer may occur in crop production areas because temperatures will no longer occur within the range, or during the critical time period for optimal growth and yield of grain or fruit. Increasing temperatures cause plants to mature and complete their stages of development faster, which may modify the feasibility and effectiveness of regional crop rotations and field management options, including double-cropping and use of cover crops.

14.2.2 Carbon dioxide (CO_2)

Recent Free-Air Carbon dioxide Enrichment (FACE) studies indicate that, at 550 ppm CO_2, yields increase under unstressed conditions by 10 to 20% over current concentrations for C_3 crops, and by 0 to 10% for C_4 crops. Crop model simulations under elevated CO_2 are consistent with these ranges. Increasing carbon dioxide (CO_2) in the atmosphere is a positive factor for plant growth, and controlled experiments have documented that elevated CO_2 concentrations can increase plant growth while decreasing soil water-use rates. The effects of elevated CO_2 on grain and fruit yield and quality, however, are mixed; reduced nitrogen and protein content observed in some nitrogen-fixing plants causes a reduction in grain and forage quality. This effect reduces the ability of pasture and rangeland to support grazing livestock. Elevated CO_2 concentrations disproportionately stimulate growth of weed species; they are likely to contribute to increased risk of crop loss from weed pressure.

In an experiment in New Delhi, it was observed increase in rice grain yield due to increase in CO_2 concentration. The increased net photosynthetic rate and greater accumulation of sugar contributed significantly to the accelerated development of leaves and tillers and finally grain yield. The impacts of elevated CO_2 should be considered among others, in the context of,

a. Changes in air temperature, particularly nocturnal temperature due to increase in CO_2 and other trace gases and changes in moisture availability and their effect on vegetative versus reproductive growth;

b. Need for more farm resources (e.g. fertilizers); and

c. Survival and distribution of pest populations, thus developing a new equilibrium between crops and pests.

14.2.3 Solar Radiation

Estimated reductions in solar radiation in agricultural areas over the last 60 years are projected to continue due to increased cloud cover and radiative scattering caused by atmospheric aerosols. Such reductions may partially offset the temperature-induced acceleration of plant growth. But changes in radiative forcing may also alternate the composition of solar radiation. It is likely that with the changes of radiation characteristics, due to the changes in pathways, will affect on germination, florescence, pollination, fruit setting and senescence behavior. The composition of solar light will have tremendous impact of photosynthesis characteristics of crop and thereby on yields.

14.2.4 Precipitation

Rainfall in India has a direct relationship with the monsoons which originate from the Indian Ocean and enters in India through two segments i.e. Bay of Bengal and Arabian Sea segments. This depends on sea level temperature at Australian coast and Peruvian coast of Pacific Ocean to have either normal circulation to cause heavy rainfall towards Peruvian coast and trigger strong water laden air circulation towards Indian sub-continent. In abnormal year monsoon is caused by a mechanism called El-Nino due to temperature reversal between these two locations of Pacific Ocean. The rise of global temperature is likely to affect this temperature regime and alter the rainfall pattern both in terms of magnitude and timing. In India more than 60% of agricultural area is rainfed. Abnormal rainfall in season and magnitude may complete change the agricultural scenario for Indian farmers. Although, agriculture contributes 14% in the Gross Domestic Product (GDP) in India, 64% of the population depends on agriculture for their livelihood. Over the years, demand for water has increased due to urbanization, increasing population, rapid industrialization and other developmental initiatives. In addition, changes in cropping and land-use patterns, over-exploitation of groundwater and changes in irrigation and drainage have modified the hydrologic cycle in many climate regions and river basins of India. Water quality and quantity are serious constraints for agriculture in most parts of India and hence, improved water management approaches must be adapted. Flood and excess rain over a short duration of time cause extensive damage to crops. Climate change has also direct bearing on water quality due to its effect on the microorganism dependent of temperature.

Higher temperature will also cause higher evapo-transpiration from the crops. In arid regions of Rajasthan an increase of 14.8 per cent in total ET demand has been projected with increase in temperature. The study further indicates that even a marginal increase in ET demand due to global warming would have a larger impact on the fragile water resources of arid zone ecosystem of Rajasthan. Therefore, change in climate will affect the soil moisture, groundwater recharge, and frequency of flood or drought, and finally groundwater level in different areas.

Optimization of water application in irrigated agriculture through precision irrigation with the objective of 'more crop per drop' may be an option. As far as possible water demanding crops should be avoided by the farmers in water scarce region. Projects like Integrated Watershed Development Project (IWDP), National Watershed Development Project for

Rainfed Areas (NWDPRA) are such project which had the objectives of water management though in-situ conservation, harvesting and artificial recharge to ground water. Warming temperatures, for instance, will act to increase crop water demand, increasing the rate of water use by crops. Crops grown on soils with a limiting soil water-holding capacity are likely to experience an increased risk of drought and potential crop failure as a result of temperature-induced increases in crop water demand, even with improved water-use efficiencies.

14.3 Extreme Conditions

Extreme weather conditions, such as floods, droughts, heat and cold waves, flash floods, cyclones and hailstorm, are direct hazards to crops. More subtle fluctuation in weather during critical phases of crop development can also have substantial impact on yields. Cultivated areas are subjected to a broader range of influences, including changes in commodity prices, costs of inputs and availability of irrigation water, etc.

14.4 Indirect Climate Change Effects

As is the case with crops and livestock, climate change affects weeds, pests, and pathogens. Changes in temperature and precipitation patterns, coupled with increasing atmospheric CO_2, create new conditions that change weed-infestation intensity, insect population levels, the incidence of pathogens, and the geographic distribution of many of these pests. Such changes on non-crop species found in agro ecosystems are indirect effects of climate change. For agriculture, such effects can alter production yields and quality, and may necessitate changes to management practices. These indirect effects may also increase farming costs, as additional inputs may be required to manage the influence of weeds, invasive species, insects, and other pests. Weeds cause the highest crop losses globally (34%), with insect pests and pathogens showing losses of 18% and 16%, respectively. By altering the environment (e.g., temperature) or increasing a resource (e.g., CO_2), we change not only the growth of an individual, but also the interactions among species, and the growth patterns of the entire plant community.

14.5 Livestock Response to Climate Change

Animal agriculture is a major component of the Indian Agricultural System. Livestock production systems are at risk to temperature stresses, particularly the lengthened exposure to extreme temperatures will further enhance production cost and productivity losses associated with all animal products, e.g., meat, eggs and milk. Changing climatic conditions affect animal agriculture in four primary ways:

a. Feed-grain production, availability and price,

b. Pastures and forage crop production and quality,

c. Animal health, growth and reproduction, and

d. Disease and pest distributions

Also, the combined effect of temperature and humidity affect animal response and are quantified through the thermal-humidity index.

14.6 Effects of Climate Change on Soil and Water

Climate change effects on agriculture also include the effects of changing climate conditions on resources of key importance to agricultural production, such as soil and water. Seasonal precipitation affects the potential amount of water available for crop production, but the actual amount of water available to plants also depends upon soil type, soil water holding capacity and infiltration rate.

Several of the processes like erosion, compaction, acidification, salinization, toxification, and net loss of organic matter which act to degrade the soil are sensitive to changing climatic conditions. The changes to the rate of soil organic matter accumulation will be affected by climate through soil temperature, soil water availability, and the amount of organic matter input from plants. Erosion is of particular concern. Changing climate will contribute to the erosivity from rainfall, snowmelt, and wind.

14.7 Adaptation Options

Adaptation strategies in use today by some farmers dealing with current changes in weather discrepancy include:

(i) Changing Cultivar Selection

(ii) Timing of Field Operations

(iii) Increased Use of Pesticides to control higher pest pressures.

(iv) New strategies for preventing rapid evolution of pest resistance to chemical control agents

(v) Development of new pesticide products

(vi) Improved pest and disease forecasting.

Adaptation options that increase the resilience of agricultural systems to increased pest pressures include:

a. Crop Diversification

b. Management of Biodiversity at both field and landscape scale to suppress pest outbreaks and pathogen transmission.

c. Altered Crops and Cropping Systems to maintain soil fertility in sustainable manner

d. Improved Management Practices, possible

e. Changes in Sowing Dates

f. Genotype Selection

Sometimes, an adaptation plan consisting of integrated changes in crop mix, irrigation techniques and methods, fertilization practices, tillage practices, and land management was found to be the most effective approach to managing climate risk. Adaptation options for managing novel crop pest management challenges may involve increased use of pesticides, new strategies for preventing rapid evolution of pest resistance to chemical control agents, the development of new pesticide products and improved pest and disease forecasting. Adaptation options that increase the resilience of agricultural systems to increased pest pressures include crop diversification and the management of biodiversity at both field and landscape scale to suppress pest outbreaks and pathogen transmission. Such adaptations help dealing in Climate Change effects in Agricultural Sector.

14.8 Crop Growth Models

Traditionally crop production systems were conducted based on agronomic observations. The disadvantage of this method was simulating the production system with changes in other system variables. Now with technological developments it has been possible to simulate scenarios through Crop growth models (James and Cutforth, 1996). These tools provide inexpensive methods to simulate cropping system to predict crop responses to varying climatic conditions and processes involved (Gowda et al 2013). It can also be effective in planning alternates resources management options (Said et al., 2019). Day and Atkin (1984) listed requirement of good crop growth model. They stated that a model should be able to understand the biological problem, it should be able to provide realistic mathematical representation of the growth phenomenon, it should be able to find quantitative solution and also be able to provide biological interpretation of results.. There are number of crop growth models available. Some of the popular models are mentioned over here.

(a) DSSAT (Decision Support System for Agrotechnology Transfer (DSSAT)

DSSAT is a computer program that can simulate crop production system of 42 crops. The tool includes database for soil, weather, crop management, data utilities and application program. It can be very effectively used for simulate growth, production and development and yield under various soil, plant and atmosphere interactions. The model also includes on-farm and precision management, regional assessments of the impact of climate variability and climate change, gene-based modeling and breeding selection, water use, greenhouse gas emissions, and long-term sustainability through the soil organic carbon and nitrogen balances (Anar et al, 2019).

(b) APSIM (The Agricultural Production System Simulator)

APSIM was developed to simulate biophysical process in farming systems, in particular where there is interest in the economic and ecological outcomes of management practice in the face of climatic risk. The model can simulate diverse range of plants, animal, soil, climate and management interactions. The model has been extensively used for variety of applications such as for on-farm decision making, farming systems design for production or resource management objectives, assessment of the value of seasonal climate forecasting, analysis of supply chain issues in agribusiness activities, development of waste management guidelines, risk assessment for government policy making and as a guide to research and education activity (Keating et al., 2003). The model has been developed by the Agricultural Production Systems Research Unit in Australia.

(c) GENECROP Model

Genecrop is a generic agrophysiological model that can simulate different processes and yield function. It can simulate the processes on daily times step at $1m^2$ area. The model can perform different processes such as crop development dynamics, generation of crop biomass and development of shooting systems (Savary and Willocquet, 2014)..

(d) WOFOST (Worlf Food Studies)

WOFOST imodel was developed by the Centre for World Food Studies in Wageningen, the Netherlands, in cooperation with with the Agricultural University and the Centre for Agrobiological Wageningen Research (CABO). The model simulates crop growth and development as a function of crop species, soil types, hydrologic conditions and weather during the growing season. The theorectical results obtained helps the user to understand the significance of light, temperature, water and macro nutrients such as nitrogen, potassium and phospohorus. The model outputs are helpful for regional and national planning (Van Diepen et al 1989).

(e) Cropsyst (Cropping System Simulation Model)

Cropsyst was developed to simulate soil water and nitrogen budgets, crop growth and development, its yield, erosion and salinity.. It is a model that can generate information on multi year, multi crop and also on daily time step to understand the impact of climate, soils and management options. The model has a integrated interface of GIS, weather generator model and other utlity program (Stöckle et al 2003).

14.9 Effect of Projected Climate Change on Crops through Crop Growth Model

14.9.1 Effect of Different Scenarios of CO_2, Temperature and Rainfall on Maize

DSSAT-CERES –Maize model was used to study the effect of induced climate variations on maize crop. Maize cultivar RCM 76 is a recommended and popular variety in NE Region was cultivated and with the help of weather parameters prevailing in Uniam region of Meghalaya was used to calibrate the model. The genetic coefficients for the maize cv. RCM 76 were also determined with standard practice as prescribed under the model. The model was then run for different scenario of climate i.e. ±1, 2 and 3 ^{0}C variation of temperature, ±19 and 50% variation of rainfall and 450 and 500 ppm of CO_2 and their combined effect on the cultivar.

Results on simulating temperature variation on the maize cv. RCM 76, with varying levels of N, revealed that increase in temperature over the current condition will be having direct negative effect on the grain yield realization maize at Umiam and the reverse is possible if temperature is decreased in future. Increase in temperature will also reduce the days to physiological maturity, means the crops will mature much before than its normal duration of maturity. This will also impact the grain yield realization. However, the biomass component, excluding the grain yield is unlikely to be affected due to change in temperature alone. The stress likely to be imposed by +1°C increase in temperature at all levels of N application will be more or less same, while with further increase in temperature is likely to double or triple the negative impacts. On the contrary, reduction in temperature will have much larger positive impact on the crop. Thorp et al(2008) and De Jonge et al (2007) reported that the higher temperatures in the future scenario caused a shorter grain-filling period, resulting in lower yields. The grain filling period is determined by accumulation of growing degree-days. Thus, if the daily temperatures are higher, the plant accumulates growing degree-days faster, and has fewer days to fill kernels. Corn adds weight to individual kernels at a very linear rate, approximately

6–8 mg/day, depending upon hybrid. Thus, yield is increased or decreased by the length of the grain-filling period, which is highly temperature dependent. Higher temperatures shorten the filling period.

Like the changes in temperature, the variation in rainfall regimes also has some impact on growth and yield parameters of maize cv. RCM 76. High amount of rainfall (+19%) during the crop growth period is likely to reduce the grain and tops weight and that is further magnified at +50% increase in rainfall. However, the total crop duration will remain unaffected at any level of nitrogen application and rainfall variation. On the other hand, reduction in rainfall by 19% at all levels of nitrogen application will have positive impact on the crop. This implies that the crop is getting excess rainfall than its requirement and excess rainfall may be responsible for wash out of soil nutrients, particularly nitrogen and affecting the crop growth and development adversely. However, at -50% level of rainfall, the crop may be subjected to water stress and reduce the magnitude of positive impacts to a meagre level. Another important observation that was found from the above information is that if nitrogenous fertilizer level is increased over the recommended dose (RD) (125% of RD) the wash out nutrient nitrogen may be effectively replenished and thereby minimizing the impact on the crop growth, development and yield realization. Similar findings were found by Thorpe *et al.* (2008) and DeJonge *et al.* (2007).

Compared to rainfall, temperature change is likely to have more profound impact on crop growth and development at enhanced level of CO_2 conc. (450 ppm). Reduction in temperature along with increase in CO_2 conc. is likely to have most significant increase in grain yield at either RD or at enhanced level of N application (125% RD), and it may be more than 20% of the current average yield. Similarly, at reduced levels of N application (75% RD or control) the increase in grain yield was found to be 10-20% of the current average yield. So, increase in CO_2 conc. accompanied by reduced environmental temperature has a direct positive relationship with grain yield of maize at Umiam, Meghalaya. Compared to grain yield, the tops weight is likely to increase with increase in temperature. The observations made at 500 ppm conc. of CO_2 and different temperature levels were more or less similar with that of at 450 ppm CO_2 level. This infers that with increase in CO_2 level up to 450 ppm would have some positive impact on grain yield of maize, if temperature is reduced but the same is not likely to increase further if CO_2 level increased up to 500 ppm. Gao and Delusca. (2016) have also reported a 40% increase in maize yield under 550 ppm CO_2 as compared to the ambient 387 ppm CO_2 concentration.

Again, soil fertility level also has some impact on yield realization under different conditions of environmental modification. Whether there was any increase or decrease in levels of temperature and rainfall, the yield realization at RD or excess of RD (125% of RD) in fertilizer N application were found to be in the higher side compared to 75% RD level and no application of fertilizer N. Another observation made in this part of the study is that CO_2 levels of 450 ppm or 500 ppm, at vegetative stage, does not influence grain yield of maize to a great extent. The observations lead to the conclusion that maize yield is affected by environmental modifications at vegetative stage and poor soil fertilization may have greater negative impact on maize growth, development and yield if there is any possible increase in temperature and rainfall in future.

14.9.2 Effect of Different Scenarios of CO_2, Temperature and Rainfall on Rice

Rainfall levels (±19%, ±50%), temperature (±1 °C), CO_2 (450 ppm)

Simulation was done at reproductive period of rice for different temperature and rainfall regimes at 450 ppm CO_2 concentration. When the change was made in temperature by -1°C with change in rainfall amount by ±19% and ±50%, there was negative impact on the grain yield of rice. The reduction in grain yield was by -2.4% to -8.0%. However, when the temperature was increased by +1°C with similar change in rainfall, there was overall increase in grain yield by 2.5-11.1% from the average grain yield.

Rainfall levels (±19%, ±50%), temperature (±1 °C), CO_2 (500 ppm)

Similarly, when the simulation was done at 500 ppm CO_2 concentration with decrease in temperature by 1 °C and ±19% and ±50% of average rainfall, there was reduction in the grain yield of rice by 1.7% to 6.2%. And when the temperature was increased by +1 °C with the same change in rainfall, grain yield was increased by 4.2-12.3%. It was also observed that the increase in grain yield when the change was made at 500 ppm CO_2 concentration was higher the increase in grain yield at 450 ppm CO_2 concentration.

Rainfall levels (±19%, ±50%), temperature (±2 °C), CO_2 (450 ppm)

At reproductive stage, when simulations were carried at -2 °C temperature and different rainfall regimes ±19% and ±50% at 450 ppm CO_2 conc., it was observed that there was great decrease in the simulated grain yield by 32% to 36% at all levels of nitrogen. However, with the increase in temperature 2 °C with similar change in rainfall (±19%, ±50%), positive changes were observed in the grain yield. Increase in simulated grain yield was by 0.6% to 6.0%.

Rainfall levels (±19%, ±50%), temperature (±2 °C), CO_2 (500 ppm)

At 500 ppm CO_2 conc., when the temperature was decreased by 2 °C with rainfall variation of ±19% and ±50%, similar amount of decrease in simulated grain yield was observed as in case of 450 ppm CO_2 level. Reduction in the simulated grain yield was observed by 30.8% to 35. Similarly, when the temperature was increased by 2 °C with the same variations in the rainfall, increase in the simulated grain yield (0.9-10.8%) were observed in all the N application levels. Increase in grain yield was slightly higher when the CO_2 level increased from 450 ppm to 500 ppm, it might be due to the increase in the photosynthesis process thereby increasing the accumulation of more photosynthates in the crops.

Rainfall levels (±19%, ±50%), temperature (±3 °C), CO_2 (450 ppm)

At reproductive stage, when the simulation was done change in temperature by 3 °C and rainfall by ±19% and ±50% at 450 ppm CO_2 conc., no grain was produced like the previous changed scenarios simulations. However, when the temperature was increased by 3 °C with ±19% and ±50% of average rainfall, there was reduction in the grain yield by 0.1% to 5.2% at all the nitrogen treatments except at 80 kg N ha^{-1} for 19% rainfall reduction and at 120 kg N ha^{-1} for 50% reduction in rainfall.

Rainfall levels (±19%, ±50%), temperature (±3 °C), CO_2 (500 ppm)

Similarly, like the previous simulation result, when the temperature was changed by -3 °C and rainfall by ±19% and ±50% at 450 ppm CO_2 conc., no grain was produced. However, when the temperature was increased by +3 °C with ±19% and ±50% of average rainfall, unlike the simulation result observed at 450 ppm, there was increase in the grain yield by 0.4% to 5.9% at all the nitrogen treatments except at 120 kg N ha^{-1} for +19% and +50% rainfall reduction and at 140 kg N ha^{-1} for ±50% reduction in rainfall.

Climate change and agriculture are interrelated processes, both of which take place on a global scale. Global warming is projected to have significant impacts on conditions affecting agriculture, including temperature, precipitation and glacial run-off. These conditions determine the carrying capacity of the biosphere to produce enough food for the human population and domesticated animals. Rising carbon dioxide levels would also have effects, both detrimental and beneficial, on crop yields. The overall effect of climate change on agriculture will depend on the balance of these effects. Crop and livestock production systems expanded across a diversity of growing conditions, responded to variations in climate and other natural resources, and to dynamic changes in agricultural knowledge, technology, markets, and, most recently, public demands for sustainable production of agricultural products. Assessment of the effects of global climate changes on agriculture might help to properly anticipate and adapt farming to maximize agricultural production.

At the same time, agriculture has been shown to produce significant effects on climate change, primarily through the production and release of greenhouse gases such as carbon dioxide, methane, and nitrous oxide, but also by altering the Earth's land cover, which can change its ability to absorb or reflect heat and light, thus contributing to radiative forcing. Land use change such as deforestation and desertification, together with use of fossil fuels, are the major anthropogenic sources of carbon dioxide; agriculture itself is the major contributor to increasing methane and nitrous oxide concentrations in earth's atmosphere.

The Inter-Governmental Panel on Climate Change (IPCC) predicts that temperatures in India are likely to rise by 3-4 degrees Celsius by the end of the 21st century. This imply that in the absence of any adaptation by farmers, such as change in cropping techniques or expansion in irrigation, agricultural incomes will fall by 12% on average, and by as much as 18% in unirrigated areas by the end of the century. IPCC has given some of the findings of the effect of climate change on the agriculture in its 4th Report. Some of the findings are given below:

- Over most land areas, it is expected to have warmer and fewer cold days and nights, warmer and more frequent hot days and nights results in increased yields in colder environments; decreased yields in warmer environments and increased insect outbreaks.
- Spells of heat waves and their frequency may increase over most land areas results in reduced yields in warmer regions due to heat stress.
- Heavy precipitation events and frequency is expected to increase over most areas results in damage to crops; soil erosion and inability to cultivate land due to water logging of soils.
- Area affected by drought will increase results in land degradation; lower yields and crop damage and failure and increased livestock deaths.
- Intense tropical cyclone activity may increase which results in damage to crops and withdraws (uprooting) of trees.
- Increased incidence of extreme high sea level (excludes tsunamis) results in salinization of irrigation water, estuaries and freshwater systems are expected.
- Global food production potential is likely to increase with increase in global average temperature up to about 3 °C, but above this it is very likely to decrease.

- Food and forestry trade is projected to increase in response to climate change, with increased food-import dependence of most developing countries like India and others.
- Future climate change is expected to affect agriculture through declining production and reductions in arable land area and food supply for fish.
- The magnitude of the growth stimulation effect of elevated CO_2 concentrations under field conditions, in conjunction with changing water and nutrient constraints, is uncertain. Because elevated CO_2 concentrations disproportionately stimulate growth of weed species, they are likely to contribute to increased risk of crop loss from weed pressure.

While many agricultural enterprises have the option to respond to climate changes by shifting crop selection, development of new cultivars in perennial specialty crops commonly requires 15 to 30 or more years, greatly limiting that sector's opportunity to adapt by shifting cultivars unless cultivars can be introduced from other areas. Crops and forage plants will continue to be subjected to increasing temperatures, increasing CO_2, and more variable water availability caused by changing precipitation patterns. These factors interact in their effect on plant growth and yield.

A balanced understanding of the consequences of management actions and genetic responses to the factors like increasing temperature, increasing CO_2 level, water availability, etc., will form the basis for more resilient production systems to climate change. Adaptation measures such as developing drought, pest, and heat stress resistance in crops and animals, diversifying crop rotations, integrating livestock with crop production systems, improving soil quality, minimizing off-farm flow of nutrients and pesticides, and other practices typically associated with sustainable agriculture are actions that may increase the capacity of the agricultural system to minimize the effects of climate change on productivity. e.g. developing drought and heat stress resistant crops will improve the ability of farmers to cope with increasing frequencies of temperature and precipitation variability.

There are some macro-adjustment possible to have a sustained effort to mitigate climate change and its effect on agriculture in India.

a) Conservation of water and efficient management of water resource particularly, the increase of water use efficiency and reduction of water footprint will enable to fight with sustained drought conditions.

b) Green energy production and uses through solar, wind and hybrid system will reduce the pressure on fossil fuel for energy production thereby reduces the carbon footprint. This will also break the nexus between water and energy.

c) Reduction of synthetic fertilizer for crop production will have positive impact. This will not only reduce overall carbon footprint in agricultural sector also reduce N_2O load in agriculture. Organic practice is definitely a healthy farming practice for sustainable soil health and agricultural productivity in the long run.

d) Zero tillage or minimum tillage are found to be good practice for carbon sequestration and minimize the oxidation of Soil Organic Carbon and helps in soil fertility built up.

e) Land management practices such as reforesting rangelands, planting hedgerows and other perennial plants serve many benefits, such as providing shelter for wildlife, beautifying farms, and attracting beneficial insects for pollination and natural pest control. On the climate front, trees, shrubs, and other woody vegetation also store carbon in their biomass, protect the soil from erosion, and conserve water.

f) Non-productive live-stock should be reduced for keeping methane emission at a lower level.

g) Pasture-Based Livestock Management contributes to soil health and microbial life, whilc hclping pcrennial grasses grow and storing water in the soil.

h) Farmland conservation contributes to preservation of local food sources, protects wildlife habitat, and promotes biodiversity, among other climate-friendly impacts.

Aforestation can be another option. Also, crop rotation helps in reducing soil erosion and increases soil fertility and crop yield as the the soil of farms is not used for only one set of nutrients. Management of Genetic Resources can also contribute to it.

India has made a National Action Plan on Climate Change which was unveiled in 2008. There are eight national missions that would form the core of the national plan. These include national missions for solar energy, enhanced energy efficiency, sustainable habitat, conserving water, sustaining the Himalayan ecosystem, a “Green India”, sustainable agriculture and strategic knowledge platform for climate change.

Most scenarios and results show that climate change will have an overall positive impact or not affect significantly on India’s agriculture until

2050. By the year 2080 when temperature increase is very large, the Indian agriculture will suffer the most. In other words, it can be say that food production is not threatened up to 2050 and does not need to import food, but by the year 2080 food production is threatened. So, we need to work on better adaptation strategies for mitigating the impacts and fight the challenge.

Refrences

Alexander KG and Miller MH1991 The effect of soil aggregate size on early growthand shoot- root ratio of maize (*Zea mays* L.), *Plant and Soil* **138:** 189-194

Allen RG and Pruitt WO 1991 FAO-24 reference evapotranspiration factors. *Journal of irrigation and drainage engineering, 117*(5), 758-773.

Allen RG, Jensen ME, Wright JL and Burman RD 1989 Operational estimates of reference evapotranspiration. *Agronomy journal, 81*(4), 650-662.

Allen RG, Pereira LS, Raes D and Smith M 1998 Crop evapotranspiration-Guidelines for computing crop water requirements-FAO Irrigation and drainage paper 56. *Fao, Rome, 300*(9), D05109.

Allen RG, Tasumi M, Morse A and Trezza R 2005 A Landsat-based energy balance and evapotranspiration model in Western US water rights regulation and planning. *Irrigation and Drainage systems, 19*(3-4), 251-268.

Anar MJ, Lin Z, Hoogenboom G., Shelia V, Batchelor WD, Teboh JM, Ostlie M, Schatz BG and Khan M 2019 Modeling growth, development and yield of Sugarbeet using DSSAT. *Agricultural Systems, 169,* pp.58-70.

Arbeláez AC and Castro LM 2007 Low flow discharges regional analysis using Wakeby distribution in an ungauged basin in Colombia (Doctoral dissertation, Colorado State University. Libraries).

Bates M and Hall AE1981 Stomatal closure with soil water depletion not associatedwith changes in bulk leaf water status,*Oecologia* **50:** 62-65

Bengough AG1997 Modeling rooting depth and soil strength in a drying soil profile. *Journal of Theoretical Biology,* **186:** 327-338

Beran MA and Rodier JA 1985 *Hydrological aspects of drought.*Unesco.

Bhalme HN and Mooley DA 1980 Large-scale drought/floods and monsoon circulation. Mon. Wea. Rev., 108, 1197-1211.

BIS 1997 Indian Standards: Recommendation for Layout, Design and Construction of Green House Structures. BIS, New Delhi, 13pp.

Bouman BAM and Tuong TP 2001 Field water management to save water and increase its productivity in irrigated lowland rice. *Agricultural water management, 49*(1), pp.11-30."Field water management to save water and increase its productivity in irrigated lowland rice." *Agricultural water management* 49, no. 1 (2001): 11-30.

Briggs LJ & McLane JW 1910 Moisture equivalent determinations and their application. *Agronomy journal* **2(1)** 138-147.

Burt CM, Clemmens AJ, Strelkoff TS, Solomon KH, Bliesner RD, Hardy LA, Howell TA and Eisenhauer DE 1997 Irrigation performance measures: efficiency and uniformity. *Journal of irrigation and drainage engineering, 123*(6), 423-442.

Cai J, Liu Y, Lei T and Pereira LS 2007 Estimating reference evapotranspiration with the FAO Penman–Monteith equation using daily weather forecast messages. *Agricultural and Forest Meteorology, 145*(1-2), 22-35.

Chiew FHS, Kamaladasa NN, Malano HM and McMahon TA 1995 Penman-Monteith, FAO-24 reference crop evapotranspiration and class-A pan data in Australia. *Agricultural Water Management, 28*(1), 9-21.

Coelho DT and Dale RF 1980 An Energy - Crop Growth Variable and Temperature Function for Predicting Corn Growth and Development: Planting to Silking 1, *Agronomy Journal* **72(3)** 503-510

Cook RJ,Sitton JW, and Waldher JT1980 Evidence for Pythium as a pathogen ofdirect-drilled wheat in the pacific northwest,*Plant Disease* **64:** 102-103

Davies WJ and Zhang J1991 Root signals and the regulation of growth and development of plants in drying soil. *Annual Review of Plant Physiology and PlantMolecular Biology* **42:** 55-76

Day W and Atkin RK eds., 2013. *Wheat growth and modelling* (Vol. 86).Springer Science & Business Media.

DeJonge KC, Kaleita AL and Thorp KR 2007 Simulating the effects of spatially variable irrigation on corn yields, costs, and revenue in Iowa. *Agricultural water management, 92*(1-2), pp.99-109.

Demetriades-Shah TH, Fuchs M, Kanemasu ET, and Flitcroft ID 1994 Further discussions on the relationship between cumulated intercepted solar radiation and crop growth, *Agricultural and Forest Meteorology* **68(3-4)**, 231-242.

Diaz HF 1983 Drought in the united states. *Journal of Applied Meteorology and Climatology, 22*(1), pp.3-16.

Doorenbos J and Pruitt WO 1977 Guidelines for predicting crop water requirements. Rome: FAO.

Droogers P and Bastiaanssen W 2002 Irrigation performance using hydrological and remote sensing modeling. *Journal of Irrigation and Drainage Engineering, 128*(1), 11-18.

Edwards DC and McKee TB 1997 Characteristics of 20th century droughts in the United States at multiple time scales. Climatology Report, 97–2, Department of Atmospheric Sciences.

Gao S and DeLuca TH 2016 Influence of biochar on soil nutrient transformations, nutrient leaching, and crop yield. *Adv. Plants Agric. Res, 4*(5), pp.1-16.

Gardner CMK,Laryea KB, and Unger PW1999 Plant physical conditions to plant growth and crop production, Food and Agriculture Organization of United Nations,Rome, AGL/MISC/24/99.

Gee GW, Newman BD, Green SR, Meissner R, Rupp H, Zhang ZF, Keller JM, Waugh WJ, Velde MV and Salazar, J. (2009). Passive wick fluxmeters: Design considerations and field applications. *Water Resources Research, 45*(4).

Gee GW, Ward AL, Caldwell TG and Ritter JC 2002 A vadose zone water fluxmeter with divergence control. *Water Resources Research, 38*(8), 16-1.

Gibbs WJ and Maher JV 1967 RAINFALL DECILES DROUGHT INDDICATORS. BUREAU OF METEOROLOGY BULLETIN 1967 , Volume - , Number 48

Glantz MH 1987 Drought and economic development in sub-Saharan Africa. *Planning for Drought: Toward a Reduction of Societal Vulnerability*, 297-316.

Gommes R and Petrassi F 1994 *Rainfall variability and drought in Sub-Saharan Africa since 1960. Agrometeoral* (No. 9).Ser. Working Paper.

Gowda NKC, Kaimal JM, Masser AE, Kang W, Friedländer MR and Andréasson C 2016 Cytosolic splice isoform of Hsp70 nucleotide exchange factor Fes1 is required for the degradation of misfolded proteins in yeast. *Molecular biology of the cell, 27*(8), pp.1210-1219.

Greco N1993 Nematode problems affecting potato production in sub-tropical climates, *Nematropica* **23:** 213-220

Gringorten II 1963 A plotting rule for extreme probability paper. *Journal of Geophysical Research, 68*(3), pp.813-814.

Gu Y, Brown JF, Verdin JP and Wardlow B 2007 A five - year analysis of MODIS NDVI and NDWI for grassland drought assessment over the central Great Plains of the United States. *Geophysical research letters, 34*(6).

Gumbel, E. J. (1963). Statistical forecast of droughts. *Hydrological sciences journal, 8*(1), 5-23.

Hao Z, AghaKouchak A, Nakhjiri N and Farahmand A 2014 Global integrated drought monitoring and prediction system. *Scientific data, 1*(1), pp.1-10.

Harmsen EW 2003 Fifty years of crop evapotranspiration studies in Puerto Rico. *Journal of soil and water conservation, 58*(4), 214-223.

Heddinghaus TR, Sahol P 1991 A review of the Palmer Drought Severity Index and Where do we go from here? In Proceedings of the seventh conference on applied climatology. American meteorological society, Boston, pp 242–246.

Heim Jr RR 2002 A review of twentieth-century drought indices used in the United States. *Bulletin of the American Meteorological Society, 83*(8), pp.1149-1166.

Hetherington AM & Woodward FI 2003. The role of stomata in sensing and driving environmental change. *Nature,* ***424*****(6951)** 901-908.

Hillel D 1980 Application of soil physics. Academic Press, New York.

Hisdal H and Tallaksen LM 2003 Estimation of regional meteorological and hydrological drought characteristics: a case study for Denmark. *Journal of Hydrology, 281*(3), 230-247.

Hollinger SE, Isard SA and Welford MR 1993 January. A new soil moisture drought index for predicting crop yields. In *Preprints, Eighth Conference on Applied Climatology* (pp. 187-190). American Meteorological Society Anaheim, CA.

Houghton JC 1978 Birth of a parent: The Wakeby distribution for modeling flood flows. *Water Resources Research, 14*(6), pp.1105-1109.

James YW, Cutforth HW 1996 Crop growth models for decision support systems, Can. J. Plant., 76, pp. 9-19

Johnson WK and Kohne RW 1993 Susceptibility of reservoirs to drought using Palmer index. *Journal of Water Resources Planning and Management, 119*(3), pp.367-387.

Jones PD, Hulme M, Briffa KR, Jones CG, Mitchell JFB and Murphy JM 1996 Summer moisture availability over Europe in the Hadley centre general circulation model based on the Palmer drought severity index. *International Journal of Climatology: A Journal of the Royal Meteorological Society, 16*(2), pp.155-172.

Jones RJ and Kerswell AP 2003 Phytotoxicity of photosystem II (PSII) herbicides to coral. *Marine Ecology Progress Series, 261*, pp.149-159.

Juhasz T and Kornfield J 1978 The Crop Moisture Index: unnatural response to changes in temperature. *Journal of applied meteorology, 17*(12), pp.1864-1866.

Karl TR 1986 The sensitivity of the Palmer Drought Severity Index and Palmer's Z-index to their calibration coefficients including potential evapotranspiration. *Journal of Climate and Applied Meteorology*, pp.77-86.

Karl TR and Knight RW 1985 Atlas of Monthly Palmer Hydrological Drought Indices (1931-1983) for the Contiguous United States. Historical Climatology Series 3-7, National Climatic Data Center, Asheville, NC.

Karl TR and Koscielny AJ 1982 Drought in the united states: 1895–1981. *Journal of Climatology, 2*(4), pp.313-329.

Karl TR and Quayle RG 1981 The 1980 summer heat wave and drought in historical perspective. *Monthly Weather Review, 109*(10), pp.2055-2073.

Keating BA, Carberry PS, Hammer GL, Probert ME, Robertson MJ, Holzworth D, Huth NI, Hargreaves JN, Meinke H, Hochman Z and McLean G 2003 An overview of APSIM, a model designed for farming systems simulation. *European journal of agronomy, 18*(3-4), pp.267-288.

Kim TW and Valdés JB 2003 Nonlinear model for drought forecasting based on a conjunction of wavelet transforms and neural networks. *Journal of Hydrologic Engineering, 8*(6), pp.319-328.

Klugman MR 1978 Drought in the upper midwest, 1931–1969. *Journal of Applied Meteorology and Climatology, 17*(10), pp.1425-1431.

Kogan F 2002 World droughts in the new millennium from AVHRR - based vegetation health indices. *Eos, Transactions American Geophysical Union, 83*(48), pp.557-563.

Kogan F 2003 Remote Sensing Contribution to Drought Mitigation. In *54th International Astronautical Congress of the International Astronautical Federation, the International Academy of Astronautics, and the International Institute of Space Law* (pp. C-1).

Kogan FN 1995 Application of vegetation index and brightness temperature for drought detection. *Advances in space research, 15*(11), pp.91-100.

Kogan FN 2001 Operational space technology for global vegetation assessment. *Bulletin of the American meteorological society, 82*(9), pp.1949-1964.

Kogan FN and Zhu X 2001 Evolution of long-term errors in NDVI time series: 1985–1999. *Advances in Space Research, 28*(1), pp.149-153.

Kogan, F.N., 1990. Remote sensing of weather impacts on vegetation in non-homogeneous areas. *International Journal of Remote Sensing, 11*(8), pp.1405-1419.

Lawson MP, Reiss A, Phillips R and Livingston K 1971 Nebraska Droughts A Study of Their Past Chronological and Spatial Extent with Implications for the Future. *Technical Research Project Completion Report,* Project No.A-018-NEB, Agreement No. 14-31-0001-3227, Nebraska Water Resources Research Institute.

Lenka NK, Lenka S, Yashona DS and Jat D 2021 Elevated temperature and low nitrogen partially offset the yield, evapotranspiration, and water use efficiency of winter wheat under carbon dioxide enrichment. *Agricultural Water Management, 250,* 106856.

Li SG, Asanuma J, Kotani A, Davaa G and Oyunbaatar D 2007 Evapotranspiration from a Mongolian steppe under grazing and its environmental constraints. *Journal of Hydrology, 333*(1), 133-143.

Liu, W.T. and Kogan, F.N., 1996. Monitoring regional drought using the vegetation condition index. *International Journal of Remote Sensing, 17*(14), pp.2761-2782.

López-Urrea R, de Santa Olalla FM, Fabeiro C and Moratalla A 2006 Testing evapotranspiration equations using lysimeter observations in a semiarid climate. *Agricultural water management, 85*(1-2), 15-26.

Masle J1998 Growth and stomatal responses of wheat seedlings to spatial and temporalvariations in soil strength of bi-layered soils, *Journal of Experimental Botany* **49:**1245-1257.

Materechera S.A., Dexter A.R. & Alston A.M. (1991) Penetration of very strong soils by seedling roots of different plant species. *Plant and Soil* 135, 31-34.

Mavi HS and Tupper G J 2004 Agrometeorology: principles and applications of climate studies in agriculture. CRC Press.

McCully, M. E. 1995. Water efflux from the surface of field-grown grass roots. Observations by cryo-scanning electron microscopy. *Physiologia Plantarum.* 95, 217-224 https://doi.org/10.1111/ j.1399-3054.1995.tb00830.x

McIsaac G 2010 Time domain reflectometry measurement of water content and electrical conductivity using a polyolefin coated TDR probe (Master's thesis, University of Waterloo).

McKee TB, Doesken NJ and Kleist J 1993 January.The relationship of drought frequency and duration to time scales.In *Proceedings of the 8th Conference on Applied Climatology* (Vol. 17, No. 22, pp. 179-183).

McMahon TA and Mein RG 1986 *River and reservoir yield* (Vol. 368). Littleton, CO: Water resources publications.

Mehran A, Mazdiyasni O and AghaKouchak A 2015 A hybrid framework for assessing socioeconomic drought: Linking climate variability, local resilience, and demand. *Journal of Geophysical Research: Atmospheres, 120*(15), pp.7520-7533.

Meissner R and Seyfarth M 2004 December, Measuring water and solute balance with new lysimeter techniques. In *3th Australian and New Zealand soil science conference,* Sydney.

Meissner R, Rupp H and Schubert M 2000 Novel lysimeter techniques—a basis for the improved investigation of water, gas, and solute transport in soils. *Journal of Plant Nutrition and Soil Science, 163*(6), 603-608.

Meissner R, Rupp H and Seyfarth M 2008 Advances in out door lysimeter techniques. *Water, Air, & Soil Pollution: Focus, 8*(2), 217-225.

Meissner R, Seeger J, Rupp H, Seyfarth M and Borg H 2007 Measurement of dew, fog, and rime with a high‐precision gravitation lysimeter. *Journal of Plant Nutrition and Soil Science, 170*(3), 335-344.

Meyer SJ, Hubbard KG and Wilhite DA 1993 A crop‐specific drought index for corn: I. Model development and validation. *Agronomy Journal, 85*(2), pp.388-395.

Mishra AK and Desai VR 2005 Drought forecasting using stochastic models. *Stochastic Environmental Research and Risk Assessment, 19*(5), pp.326-339.

Mishra AK and Singh VP 2010 A review of drought concepts. *Journal of hydrology, 391*(1-2), 202-216.

Mohan S and Arumugam N 1996 Relative importance of meteorological variables in evapotranspiration: Factor analysis approach. *Water Resources Management, 10*(1), 1-20.

Monteith JL 1977 Climate and the efficiency of crop production in Britain, *Philosophical Transactions of the Royal Society of London. B, Biological Sciences* **281** (980), 277-294.

Nalbantis I 2008 Evaluation of a hydrological drought index. *European Water, 23*(24), pp.67-77.

Narasimhan B and Srinivasan R 2005 Development and evaluation of Soil Moisture Deficit Index (SMDI) and Evapotranspiration Deficit Index (ETDI) for agricultural drought monitoring. *Agricultural and forest meteorology, 133*(1-4), pp.69-88.

NRCS, 2004. Soil Quality - Soil Biology Technical Note No. 4. USDA. http://soils.usda.gov/sqi

Ono S & Kondo S 1960 Molecular theory of surface tension in liquids. In *Structure of Liquids/Struktur der Flüssigkeiten* (pp. 134-280) Springer, Berlin, Heidelberg.

Oosterbaan RJ 1994 Frequency and regression analysis of hydrologic data. *Drainage principles and applications, Publication, 16*, pp.175-223.

Özger M, Mishra AK and Singh VP 2009 Low frequency drought variability associated with climate indices. *Journal of Hydrology, 364*(1-2), pp.152-162.

Pai DS, Sridhar L, Guhathakurta P, Hatwar HR 2011 District-wide drought climatology of the southwest monsoon season over India based on standardized precipitation index (SPI). *Natural hazards, 59*(3), pp.1797-1813.

Palmer WC 1965 Meteorological drought. Weather Bureau Research Paper No. 45, U. S. Dept. of Commerce, Washington, DC, 58 pp.

Palmer WC 1968 Keeping track of crop moisture conditions nationwide: the new crop moisture index. Weatherwise 21, 156–161.

Palmer, W.C., 1965. *Meteorological drought* (Vol. 30). US Department of Commerce, Weather Bureau.

Palmer, W.C., 1968. Keeping track of crop moisture conditions, nationwide: the new crop moisture index.*Weatherwise, 21(4) 156-161.*

Pashiardis S and Michaelides S 2008 Implementation of the standardized precipitation index (SPI) and the reconnaissance drought index (RDI) for regional drought assessment: a case study for Cyprus. *European Water, 23*(24), 57-65.

Passioura J.B. (1988) Root signals control leaf expansion in wheat seedlings growing in drying soil. *Australian Journal of Plant Physiology* 15, 687- 693.

Payen, D and Gregory, P. J. 1988. Russell's soil conditions and plant growth. 11th Edn. *pp. 282-297*

Penman HL 1948 Natural evaporation from open water, bare soil and grass. *Proceedings of the Royal Society of London. Series A. Mathematical and Physical Sciences, 193*(1032), 120-145.

Pereira AR and Pruitt WO 2004 Adaptation of the Thornthwaite scheme for estimating daily reference evapotranspiration. *Agricultural Water Management, 66*(3), 251-257.

Peters AJ, Walter-Shea, EA, Ji L, Vina A, Hayes M and Svoboda MD 2002 Drought monitoring with NDVI-based standardized vegetation index. *Photogrammetric engineering and remote sensing, 68*(1), pp.71-75.

Ponce P, Molina A, Cepeda P, Lugo E. and MacCleery B 2014 *Greenhouse design and control.* CRC Press.

Rao AR and Padmanabhan G 1984 Analysis and modeling of Palmer's drought index series. *Journal of hydrology, 68*(1-4), pp.211-229.

Reed BC, Brown JF, VanderZee D, Loveland TR, Merchant JW and Ohlen DO 1994 Measuring phenological variability from satellite imagery. *Journal of vegetation science, 5*(5), pp.703-714.

Roberts Keith, ed. *Handbook of plant science.* Vol. 1. John Wiley & Sons, 2007. p 185, ISBN 9780470057230.

Rossi G 2000 Drought mitigation measures: a comprehensive framework. In *Drought and drought mitigation in Europe* (pp. 233-246). Springer, Dordrecht.

Rost, Thomas L.; Michael Barbour; C. Ralph Stocking; Terence Murphy (2006). Plant Biology, 2nd Edition. Thompson, Brooks/Cole. ISBN 978-0-534-38061-8.

Savary S and Willocquet L 2014 Simulation modeling in botanical epidemiology and crop loss analysis. *Plant Health Instructor,* (Online), p.147.

Schneider SH 1996 December.Engineering change in global climate.In *Forum for Applied Research and Public Policy* (Vol. 11, No. 2).

Sentelhas PC, Gillespie TJ and Santos EA 2010 Evaluation of FAO Penman-Monteith and alternative methods for estimating reference evapotranspiration with missing data in Southern Ontario, Canada. *Agricultural Water Management, 97*(5), 635-644.

Shafer BA and Dezman LE 1982 Development of a Surface Water Supply Index (SWSI) to assess the severity of drought conditions in snowpack runoff areas. In: Proceedings of the Western Snow Conference, Reno, NV, 19-23, April 1982, pp. 164-175. Colorado State University, Fort Collins, Colorado.

Shafer BA and Dezman LE 1982 January.DEVELOPMENT OF SURFACE WATER SUPPLY INDEX (SWSI) TO ASSESS THE SEVERITY OF DROUGHT CONDITION IN SNOWPACK RUNOFF AREAS. PROCEEDING OF THE WESTERN SNOW CONFERENCE.

Shibu ME, Leffelaar PA, Van Keulen H and Aggarwal PK 2010 LINTUL3, a simulation model for nitrogen-limited situations: Application to rice. *European Journal of Agronomy, 32*(4), pp.255-271.

Simpson RS 2003 Lighting control—technology and applications. Focal Press, Oxford.

Sirdas S and Sen Z 2003 Spatio-temporal drought analysis in the Trakya region, Turkey.Hydrol. Sci. J. 48(5), 809–820.

Skierucha W 2000 Accuracy of soil moisture measurement by TDR technique. *International agrophysics, 14*(4).

Spitters CJT, Schapendonk AHCM 1990 Evaluation of breeding strategies for drought tolerance in potato by means of crop growth simulation. Plant and Soil123:193-203.

Stöckle CO, Donatelli M and Nelson R 2003 CropSyst, a cropping systems simulation model. *European journal of agronomy, 18*(3-4), pp.289-307.

Stöckle CO, Kjelgaard J and Bellocchi G 2004 Evaluation of estimated weather data for calculating Penman-Monteith reference crop evapotranspiration. *Irrigation science, 23*(1), 39-46.

Su, X. Tsai CL, Wang H, Nickerson DM and Li B 2009 Subgroup analysis via recursive partitioning. *Journal of Machine Learning Research, 10*(2).

Thom HCS 1958 A note on the gamma distribution. Monthly Weather Review, 86:117-122.

Thorp KR, DeJonge KC, Kaleita AL, Batchelor WD and Paz JO 2008 Methodology for the use of DSSAT models for precision agriculture decision support. *Computers and electronics in agriculture, 64*(2), pp.276-285.

Tigkas D 2008 Drought characterisation and monitoring in regions of Greece. *European Water, 23*(24), pp.29-39.

Topp GC, Davis JL and Annan AP 1980 Electromagnetic determination of soil water content: measurements in coaxial transmission lines. *Water Resources Res.* 16: 574-582.

Tsakiris G 2007 Practical application of risk and hazard concepts in proactive planning. *European Water, 19*(20), pp.47-56.

Tsakiris G and Vangelis HJEW 2005 Establishing a drought index incorporating evapotranspiration. *European water, 9*(10), pp.3-11.

Tsakiris G, Nalbantis I, Pangalou D, Tigkas D and Vangelis H 2008 June.Drought meteorological monitoring network design for the reconnaissance drought index (RDI). In *Proceedings of the 1st International Conference "Drought management: scientific and technological innovations". Zaragoza, Spain: option Méditerranéennes, series A* (No. 80, p. 2008).

Tuller M Or D & Hillel D 2004 Retention of water in soil and the soil water characteristic curve. *Encyclopedia of Soils in the Environment*, **4,** 278-289.

Tyagi NK, Sharma DK and Luthra SK 2000 Determination of evapotranspiration and crop coefficients of rice and sunflower with lysimeter. *Agricultural water management, 45*(1), 41-54.

Van Diepen CV, Wolf J, Van Keulen H and Rappoldt C 1989 WOFOST: a simulation model of crop production. *Soil use and management, 5*(1), pp.16-24.

van Ittersum MK, Leffelaar PA, van Keulen H, Kropff MJ, Bastiaans L and Goudriaan J 2003 On approaches and applications of the Wageningen crop models. *European journal of agronomy, 18*(3-4), pp.201-234.

Vengadaramana A & Jashothan PTJ 2012. Effect of organic fertilizers on the water holding capacity of soil in different terrains of Jaffna peninsula in Sri Lanka. *J. Nat. Prod. Plant Resour,* **2(4)**, 500-503.

Volkmar K.M. (1996) Effects of biopores on the growth and N-uptake of wheat at three levels of soil moisture. *Canadian Journal of Soil Science* 76, 453- 458.

Weghorst, K.M., 1996. The Reclamation Drought Index: guidelines and practical applications. *Bureau of Reclamation, Denver, CO, 6.*

Werick, W.J., Willeke, G.E., Guttman, N.B., Hosking, J.R.M. and Wallis, J.R., 1994. National drought atlas developed.

White DH and O'Meagher B 1995 Coping with exceptional droughts in Australia. *Drought Network News (1994-2001),* p.91.

Wilhite DA and Glantz MH 1985 Understanding: the drought phenomenon: the role of definitions. *Water international, 10*(3), 111-120.

Willeke G, Hosking JRM, Wallis J R, and Guttman NB 1994 The National Drought Atlas. Institute for Water Resources Report 94-NDS-4, U.S. Army Corps of Engineers.

Wolfe D.W., Topoleski D.T., Gundersheim N.A. & Ingall B.A. (1995) Growth and yield sensitivity of four vegetable crops to soil compaction. *Journal of the American Society for Horticultural Science* 120, 956- 963.

X, and van Laa HH 2005 Crop Systems Dynamics. An ecophysiological simulation model for genotype-by-environment interactions. Wageningen Academic Publishers, Wageningen, The Netherlands, 153 pp.

Yang L, Wylie BK, Tieszen LL and Reed BC 1998 An analysis of relationships among climate forcing and time-integrated NDVI of grasslands over the US northern and central Great Plains. *Remote Sensing of Environment, 65*(1), pp.25-37.

Yue S, Ouarda TB, Bobée B, Legendre P and Bruneau P 1999 The Gumbel mixed model for flood frequency analysis. *Journal of hydrology, 226*(1-2), pp.88-100.

Yusof F and Hui-Mean F 2012 Use of statistical distribution for drought analysis. *Applied Mathematical Sciences, 6*(21), pp.1031-1051.

Abbreviation

%	percentage
°C	degree Celsius
ACA	Ammoniacal copper arsenate
Al_3^+	Aluminium ion
APSIM	The Agricultural Production System Simulator
AVHRR	Advanced Very High Resolution Radiometer
AW	Available water
B	Boron
BMDI	Bhalme and Mooly Drought Index
C	Carbon
Ca	Calcium
Ca_2^+	Calcium
CABO	Centre for Agrobiological Wageningen Research
CCA	Chromate copper arsenate
CEC	Cation Exchange Capacity
Cl	Chlorine
cm	centimetre
CMI	Crop Moisture Index
cmol	centimole
Co	Cobalt
CO_2	Carbon di oxide
CSDI	Crop Specific Drought Index
Cu	Copper
DDMP	Daily Dry Matter Production
DSSAT	Decision Support System for Agrotechnology Transfer
e.g.	example
EC	Electrical conductivity
ET	Evapo-transpiration
EVA	Ethylene Vinyl Acetate
FACE	Free-Air Carbon dioxide Enrichment
FAO	Food and Agricultural Division
FAO	Food and Agricultural Organization
Fe	Ferric
Fig.	Figure
FRP	*Fibre glass Reinforced Plastic panels*
GCMs	Global circulation Models
GDD	Growing degree days
H	Hydrogen
H^+	Hydrogen ion
HID	High intensity discharge
HTU	Helio Thermal Unit
IBSNAT	International Benchmark Sites Network for Agrotechnology Transfer project
IPCC	Intergovernmental Panel on Climate Change
IRGA	Infrared gas analysers
IWDP	Integrated Watershed Development Project
K	Kelvin
K	Potash
K^+	Potassium ion
Kg	Kilogram
kPa	Kilo pascal

LAI	*Leaf Area Index*
LECA	Lightweight Expanded Clay Aggregate
LED	Light emitting diode
m^3	cubic meter
meq	Milli equivalent
Mg	Magnesium
Mg_2^+	Magnesium ion
MJ	megajoules
mm	millimetre
Mn	Mangenese
Mo	Molybdenum
MPa	Mega Pascal
N	Nitrogen
Na	Sodium
Na^+	Sodium ion
NDVI	Normalized Difference vegetation Index
NFT	Nutrient Film Technique
Ni	Nickel
nm	nano meter
NOAA	National Oceanic and Atmospheric Administration
NRCS	Natural Resources Conservation Service
NRI	National Rainfall Index
NWDPRA	National Watershed Development Project for Rainfed Areas
O	Oxygen
P	Phosphorous
PAR	Photo synthetically Active Radiation
PAR	Photosynthetically active radiation
PDSI	Palmer Drought Severity Index
PET	Potential Evapo-transpiration
PHDI	Palmer Hydrological Drought Index
PNP	Percentage of Normal Precipitation
ppm	parts per million
PVC	Polyvinyl Chloride
RCMs	Regional Climate Models
RCPs	Representative Concentration Pathways
RD	Recommended dose
RDI	Reclamation Drought Index
RDI	Reclamation Drought Index
RDI	Reconnaissance Drought Index
RH	Relative humidity
RUE	Radiation use efficiency
S	Sulphur
SMDI	Soil Moisture Drought Index
SPAC	Soil-Plant-Atmospheric Continuum
SPI	Standardized Precipitation Index
SWSI	Surface Water Supply Index
T	Temperature
TCI	Temperature Condition Index
TDMP	Total Dry Matter Production
TDR	Time domain Reflectometry
US	United States
UV	Ultra Violet
VCI	Vegetation Condition Index
Vs	versus
W/m^2	watt per square meter
WMO	World Meteorological Organisation
WOFOST	Worlf Food Studies
Zn	Zinc
µm	micro mete

Subject Index